Clinical Mass Spectrometry

Guest Editors

NIGEL J. CLARKE, MD, PhD
ANDREW N. HOOFNAGLE, MD, PhD

CLINICS IN LABORATORY MEDICINE

www.labmed.theclinics.com

Consulting Editor
ALAN WELLS, MD, DMSc

September 2011 • Volume 31 • Number 3

SAUNDERS an imprint of ELSEVIER, Inc.

W.B. SAUNDERS COMPANY
A Division of Elsevier Inc.

1600 John F. Kennedy Boulevard • Suite 1800 • Philadelphia, Pennsylvania 19103-2899

http://www.theclinics.com

CLINICS IN LABORATORY MEDICINE Volume 31, Number 3
September 2011 ISSN 0272-2712, ISBN-13: 978-1-4557-1025-6

Editor: Katie Hartner
Developmental Editor: Donald Mumford

Reprints. For copies of 100 or more, of articles in this publication, please contact the Commercial Reprints Department, Elsevier Inc., 360 Park Avenue South, New York, New York 10010-1710. Tel. (212) 633-3813, Fax: (212) 462-1935, E-mail: reprints@elsevier.com.

Clinics in Laboratory Medicine (ISSN 0272-2712) is published quarterly by Elsevier Inc., 360 Park Avenue South, New York, NY 10010-1710. Months of issue are March, June, September, and December. Business and Editorial offices: 1600 John F. Kennedy Blvd., Suite 1800, Philadelphia, PA 19103-2899. Periodicals postage paid at NewYork, NY and additional mailing offices. Subscription prices are $225.00 per year (US individuals), $364.00 per year(US institutions), $120.00(US students), $273.00 per year (Canadian individuals), $460.00 per year (foreign institutions), $165.00 (foreign students). Foreign air speed delivery is included in all *Clinics* subscription prices. All prices are subject to change without notice. POSTMASTER: Send address changes to *Clinics in Laboratory Medicine*, Elsevier Health Sciences Division, Subscription Customer Service, 3251 Riverport Lane, Maryland Heights, MO 63043. **Customer Service: 1-800-654-2452 (US). From outside of the US and Canada, call 1-314-447-8871. Fax: 1-314-447-8029. E-mail: journalscustomerservice-usa@elsevier.com (for print support) or journalsonlinesupport-usa@elsevier.com (for online support).**

Clinics in Laboratory Medicine is covered in EMBASE/Exerpta Medica, MEDLINE/PubMed (Index Medicus), Cinahl, Current Contents/Clinical Medicine, BIOSIS and ISI/BIOMED.

Printed and bound by CPI Group (UK) Ltd, Croydon, CR0 4YY

Transferred to Digital Print 2011

Contributors

CONSULTING EDITOR

ALAN WELLS, MD, DMSc
Department of Pathology, University of Pittsburgh, Pittsburgh, Pennsylvania

GUEST EDITORS

NIGEL J. CLARKE, MD, PhD
Director of Mass Spectrometry; Scientific Director (R&D) Steroids Department, Quest Diagnostics-Nichols Institute, San Juan Capistrano, California

ANDREW N. HOOFNAGLE, MD, PhD
Assistant Professor, Department of Laboratory Medicine, University of Washington Seattle, WA

AUTHORS

EMILY S. BOJA, PhD
Program Manager, Office of Cancer Clinical Proteomics Research, National Cancer Institute, National Institutes of Health, Bethesda, Maryland

CHRISTOPH H. BORCHERS, PhD
University of Victoria – Genome BC Proteomics Centre; Department of Biochemistry and Microbiology, University of Victoria, Victoria, British Columbia, Canada

CORY E. BYSTROM, PhD
Quest Diagnostics–Nichols Institute, Research and Development, San Juan Capistrano, California

JOHNSON-DAVIS, PhD, DABCC
Assistant Professor, Department of Pathology, University of Utah School of Medicine and ARUP Laboratories, Salt Lake City, Utah

DOMINIK DOMANSKI, PhD
University of Victoria and University of Victoria – Genome BC Proteomics Centre, Victoria, British Columbia, Canada

GABRIELA COHEN FREUE, PhD
Department of Statistics, University of British Columbia (UBC); NCE CECR Centre of Excellence for the Prevention of Organ Failure, Vancouver, British Columbia, Canada

RUSSELL P. GRANT, PhD
Research and Development, Laboratory Corporation of America, Burlington, North Carolina

John S. Hill, PhD
Department of Pathology and Laboratory Medicine, University of British Columbia, Vancouver, British Columbia, Canada

ANGELA M. JACKSON, MS
University of Victoria and University of Victoria – Genome BC Proteomics Centre, Victoria, British Columbia, Canada

CHRISTINE A. MILLER, MS
Agilent Technologies, Santa Clara, California

CAROL E. PARKER, PhD
University of Victoria and University of Victoria – Genome BC Proteomics Centre, Victoria, British Columbia, Canada

AMANDA G. PAULOVICH, MD, PhD
Associate Member, Fred Hutchinson Cancer Research Center, Seattle, Washington

ALAN L. ROCKWOOD, PhD, DABCC
Associate Professor, Department of Pathology, University of Utah School of Medicine and ARUP Laboratories, Salt Lake City, Utah

HENRY RODRIGUEZ, PhD, MBA
Director, Office of Cancer Clinical Proteomics Research, National Cancer Institute, National Institutes of Health, Bethesda, Maryland

DEREK S. SMITH, BS
University of Victoria and University of Victoria – Genome BC Proteomics Centre, Victoria, British Columbia, Canada

JEFFREY R. WHITEAKER, PhD
Director of Proteomics, Paulovich Laboratory, Fred Hutchinson Cancer Research Center, Seattle, Washington

YANAN YANG, PhD
Agilent Technologies, Santa Clara, California

Contents

> While affinity-based enrichment may be necessary for low-abundance
> analytes, we have found that, for the analysis of high-to-moderate abun-
> dance plasma proteins, it is often possible to omit the immunoprecipitation
> step, and analyze the plasma directly. Using multiple-reaction-monitoring
> with stable-isotope-labeled internal standard peptides, it is possible to
> accurately and reproducibly quantitate over one hundred peptides using
> nearly 1000 precursor/product ion transitions in a single multiplexed
> analysis. These multiplexed assays can also be used as panels for
> biomarker discovery. The smaller group of biomarker proteins discovered
> in this manner can become the basis for a new MRM-based assay with
> even higher throughput.

> Quantitative protein assays are needed in a wide range of biological
> studies. Traditional immunoassays are not available for a large number
> of proteins, and development of new immunoassays requires a signif-
> icant investment of time and money. The development of assays using
> peptide immunoaffinity enrichment coupled with targeted mass spec-
> trometry has many advantages including versatility in design, ease of
> use, enhanced specificity, and good performance characteristics. This
> review presents recent developments in the characterization and im-
> plementation of immuno-selected reaction monitoring assays.

> The development of mass spectrometry instrumentation, software, and
> informatics tools in support of proteomics research has set the stage for
> protein and peptide analysis to emerge in the clinical lab. This transition is
> aided by an emphasis on translational research wherein one aim is the
> focused conversion of laboratory discoveries into useful clinical tools.
> Mass spectrometry is now being driven into service in the clinical labora-

tory as both a research tool and analytical platform on which accurate and precise diagnostic measurements of proteins and peptides can be made.

Mass spectrometry is rapidly expanding its role in clinical chemistry. Historically, it has been used primarily for small molecule analysis, but now the field seems set for a rapid growth in protein and peptide analysis. Small molecule analysis applications can be grouped roughly into the analysis of endogenous compounds and the analysis of exogenous compounds. The analysis of exogenous compounds is largely the domain of toxicology, and this article focuses on the following 2 toxicology topics: therapeutic drug monitoring and trace element analysis.

This article describes the challenges and solutions that have been realized during the last 10 years by a number of clinical laboratories that have embraced automation and liquid chromatography-tandem mass spectrometry (LC-MS/MS) technologies for clinical utility. Current state-of-the-art liquid handling-based sample preparation, multiplexed chromatographic systems, and automated data review technologies are described, facilitating more than 2500 samples per system per day. Emerging technologies are discussed that may fundamentally drive the application of systems solutions for mass spectrometry to a broader clinical setting.

Clinical proteomics provides valuable information to the study of diseases at the molecular level, with the potential to discover biomarkers for disease states. The discovery of protein/peptide signatures "leaked" by cancerous tumors into clinically accessible fluids could possibly lead to developing quantitative assays for diagnosing cancer early. Despite having over 1,200 cancer-related protein biomarker candidates published in the scientific literature,[1] the rate of introduction of new protein biomarkers to market as approved by FDA has averaged 1.5 new proteins per year for 15 years. This discrepancy points to an ineffective translation of proteomics from the bench to the bedside.

THE CLINICS ARE NOW AVAILABLE ONLINE!

Access your subscription at:
www.theclinics.com

Preface

Mass Spectrometry Continues Its March into the Clinical Laboratory

In cartoons, ants often make the sound of battalions of troops walking along deserted roads into battle. If we listen carefully, we might be able hear the same sounds made by mass spectrometers in clinical labs around the world. Indeed, mass spectrometry has changed the face of laboratory medicine and our ability to care for our patients. The method is favored among toxicology and endocrinology laboratories, especially given its long history in the measurement of small molecules. However, several clinical laboratories are beginning to use their mass spectrometers to quantitate peptides and proteins and to detect and identify microorganisms. In many laboratories, mass spectrometers have replaced immunoassay analyzers—these are generally larger laboratories, but some smaller laboratories are also beginning to invest in the technology and to use it to replace problematic immunoassays. Even though many would consider that the invasion of mass spectrometers has been going on for decades, others would say that it is just beginning. While perspective is everything, one thing is for certain—the ants are marching to a clinical lab near you.

This collection of articles is designed to give a very broad overview regarding the current state of mass spectrometry in the diagnostics arena. As such, it is both retrospective in the sense of "where did this all start" and at the same time very much forward looking—the "are we there yet?" phase.

Mass spectrometry in one form or another has been around for more than 100 years and in that time it has become very close to the "universal detector." Initially the instruments and their modes of operations were not compatible with biological matrices and analytes. However, due to the dedicated work of such visionaries as Nobel Laureate John Fenn in the 1980s, this changed and the age of bio-analysis dawned on the field of mass spectrometry. Over the next two decades, liquid chromatography coupled initially with mass spectrometry (LC-MS) and later tandem mass spectrometry (LC-MS/MS) assumed center stage in the world of quantitative biological analysis.

LC-MS/MS quantitative analysis was first deployed in pharmaceutical laboratories—mainly in support of drug discovery, development, and regulatory submission. Soon a small number of large diagnostic reference laboratories began to look into the use of LC-MS/MS instrumentation for commercial use of certain therapeutic drug-monitoring assays. These assays (such as for tacrolimus and cyclosporine) were "homebrew" tests, ie, laboratory developed. The expertise needed to develop the tests coupled with the expense and complexity of the instrumentation precluded all but the largest diagnostic labs from implementing mass spectrometry. Several of these laboratories forged ahead and rapidly realized the potential this methodology had to offer. Not only was it automated (compared to manual radioimmunoassays, etc), but also it provided superior specificity than most antibody-based tests. This was due to the absolute measurement of the analyte rather than measurement of a

Clin Lab Med 31 (2011) ix–xi
doi:10.1016/j.cll.2011.07.009
0272-2712/11/$ – see front matter
labmed.theclinics.com

surrogate such as release of a radioactive tracer or flash of light. By not only measuring the mass of the analyte but also being able to provide structural confirmation that it truly was the analyte of interest and not a close homolog, the power of mass spectrometry was discovered among diagnostic scientists.

The first assays to be offered clinically were for the measurement of small molecules in plasma, sera, and urine—both exogenous (eg, therapeutic drug monitoring discussed in the article by Rockwood and Davis) and endogenous (eg, steroids discussed in the article by Russ Grant). More recently, proteins and peptides have become an intense area of interest, particularly given the many recognized issues with immunoassays performed on human samples (discussed by Cory Bystrom, Jeff Whiteaker, and Christoph Borchers, et al.). Indeed, the first mass spectrometric assays for peptides and proteins are now being offered clinically at some sites.

Since the introduction of soft ionization techniques near the end of the millennium, vendors have made their instruments smaller, simpler, more robust, and (yes) cheaper. In the meantime diagnostic laboratories have gone ahead full bore in their adoption of the technology and developed hundreds of assays utilizing the methodology. Furthermore, some professional bodies are now urging their members to use LC-MS and LC-MS/MS assay preferentially over immunochemiluminescence or radioimmunoassay due to the high specificity and precision of MS data. All of this has led to a large amount of interest from regulatory bodies. The background to this and the need to develop a dialogue between the laboratories and the regulatory entities are discussed in the article by Henry Rodriguez and coworkers.

The implementation of mass spectrometric methods in the clinical laboratory is still complex. Although many steps in the workflow can be automated and the process can begin to resemble an autoanalyzer in some cases, the signal from mass spectrometric assays is not a single channel as it is in chemistry and immunoassay analyzers. This is the foundation of the improved specificity of the platform, but it places new burdens on the laboratory in terms of quality control: the quality control of the batch process (standard) and the quality control of individual samples (not possible on other platforms). With the aid of spiked internal standards, interfering substances and ex vivo analyte degradation (as discussed by Cory Bystrom) can be identified and monitored on a sample-by-sample basis. The article by Russ Grant discusses software development efforts aimed at simplifying the analysis of quality control data at both the batch process and the individual sample level, which can be cumbersome in mass spectrometric methods deployed in the clinical laboratory.

To discuss the mounting interest in the mass spectrometric quantitation of peptides and proteins in clinical research and clinical care, Cory Bystrom gives an overview of proteomics, its history, and current approaches. The quantitation of proteins may require affinity enrichment methods, which are discussed in some detail by Jeff Whiteaker. In some cases, proteins can be measured directly in multiplexed assays without affinity enrichment, which is discussed in more detail by Borchers and colleagues. For some small proteins LC-MS or LC-MS/MS will likely become the reference method in the near future. For large proteins proteolytic digestion is required for quantitation and can be extremely variable from day to day and sample to sample. Basic researchers have often turned a blind eye to the problem, which is untenable for clinical care. It is expected that innovations in automation, calibration, and quality control will overcome the inherent issues associated with variability of proteolytic digestion. Until that point, quantitation of large proteins will remain in the research realm.

Hopefully this edition of the *Clinics in Laboratory Medicine* will give the reader an overall flavor for where mass spectrometry sits in the world of the diagnostic laboratory, some idea of the strengths and weakness of the technique, and a view into where mass spectrometers will be marching in future years.

Nigel J. Clarke, MD
Steroids Department
Quest Diagnostics-Nichols Institute
San Juan Capistrano, CA

Andrew N. Hoofnagle, MD, PhD
Department of Laboratory Medicine
University of Washington
Seattle, WA

E-mail addresses:
nigel.j.clarke@questdiagnostics.com (N.J. Clarke)
ahoof@u.washington.edu (A.N. Hoofnagle)

High-Flow Multiplexed MRM-Based Analysis of Proteins in Human Plasma Without Depletion or Enrichment

Dominik Domanski, PhD[a], Derek S. Smith, BS[a],
Christine A. Miller, MS[b], Yanan Yang, PhD[b],
Angela M. Jackson, MS[a], Gabriela Cohen Freue, PhD[c,d],
John S. Hill, PhD[e], Carol E. Parker, PhD[a],
Christoph H. Borchers, PhD[a,f,*]

KEYWORDS

- Multiple reaction monitoring • Multiplexing
- Biomarker discovery • Plasma proteins
- Quantitative proteomics

Jeff Whiteaker's chapter within this issue describes the use of stable isotope standards and capture by anti-peptide antibodies (SISCAPA), an analytical technique that combines immunopurification with co-capture of isotopically-labeled standards, elution, and multiple reaction monitoring (MRM) analysis. While affinity-based enrichment may be necessary for low abundance analytes, in our laboratory we have found that, for the analysis of high-to-moderate abundance plasma proteins, it is often possible to omit the immunoprecipitation step, and analyze the plasma directly. SISCAPA has been demonstrated for approximately 50 proteins per assay, but may have a practical multiplexing limit due to antibody amounts and volume limitations.

[a] University of Victoria-Genome BC Proteomics Centre, Vancouver Island Technology Park, #3101-4464 Markham Street, Victoria, BC V8Z 7X8, Canada
[b] Agilent Technologies, 5301 Stevens Creek Boulevard, Santa Clara, CA 95051, USA
[c] Department of Statistics, University of British Columbia, Vancouver, BC V6Z 1Y6, Canada
[d] NCE CECR Centre of Excellence for the Prevention of Organ Failure, Vancouver, BC V6Z 1Y6, Canada
[e] Department of Pathology and Laboratory Medicine, University of British Columbia, Vancouver, BC V6Z 1Y6, Canada
[f] Department of Biochemistry and Microbiology, University of Victoria, Vancouver, BC, Canada
* Corresponding author.
E-mail address: christoph@proteincentre.com

Clin Lab Med 31 (2011) 371–384
doi:10.1016/j.cll.2011.07.005
0272-2712/11/$ – see front matter © 2011 Published by Elsevier Inc.

labmed.theclinics.com

Plasma is perhaps the biofluid of choice, as it is easily obtained and stored, and it already forms the basis of many biomedical screening analyses for cholesterol, diabetes, and other diseases. Plasma, however, is a difficult matrix, as the dynamic range of protein concentrations exceeds 10^{10}.[1] For "shotgun" proteomics, one common approach for "digging deeper" into the proteome is to remove the most abundant proteins (usually the 12 or 14 most-abundant proteins). However, this wide dynamic range means that, even if the depletion is successful, there is still a dynamic range of concentrations of 10^9 between the low and high abundance proteins. Moreover, depletion or affinity purification is expensive, and adding a depletion or immunopurification step (followed by elution) also reduces the overall throughput of the assay. In addition, there is always the question of the reproducibility of the depletion step, as well as the possibility of removing proteins of interest along with the depleted proteins.

For methods that use affinity purification, one needs to know the specific protein targets and have antibodies against each one. For screening for very large numbers of protein targets, however, this approach is not feasible. Moreover, for clinical analyses, many of the abundant plasma proteins are actually correlated with cardiovascular disease,[2] so removing these proteins might remove valuable information from the sample, including proteins which might be important for population-wide screening.

Our laboratory, therefore, became interested in how deep it was possible to go into the plasma proteome without depletion. This approach is not without challenges. Because of the complexity of plasma as a matrix, the use of isotopically labeled standards is essential for accurate quantitation. Because of the challenges in quantitating proteins, quantitation of proteins is actually based on the quantitation of proteotypic peptides. Reproducible generation of peptides is based on efficient and (most importantly) reproducible protein digestion.[3] The complexity of the matrix means that the dynamic range of the proteins is now reflected in the dynamic range of the peptides, but is now increased by an average of 10 to 20 peptides per protein. The success and accuracy of the MRM analysis depends on the both the resolution of the mass spectrometer and the resolution of the online high-pressure liquid chromatography (HPLC) system. All of these factors will be discussed later in this chapter.

SAMPLE COLLECTION

A reproducible assay for plasma studies starts with reproducible plasma collection procedures. Studies have shown that, for proteomics studies, plasma is preferred to serum. Both plasma and serum are dynamic matrices, as they contain proteases that will affect the sample, and these processes start immediately upon collection. The preparation of serum from plasma requires coagulation. While coagulation removes fibrinogen, and should simplify the sample, this process is inherently non-reproducible. Even collection-tube manufacturers have been unable to produce a serum collection tube that provides reproducibility from day to day, much less from laboratory to laboratory.

Plasma, on the other hand has been found to be stable and reproducible, but only if collected directly into tubes that already contain protease inhibitors (either a protease cocktail[4] or EDTA[5]). In the past, unfortunately, it seems that the medical profession has had a preference for serum, and most "banked" samples are still serum. Now that the Human Proteome Organization has recommended plasma for proteomics studies,[6] it is hoped that this will change. It is important to remember, however, that the data is only as good as the "weakest link," and collection using the proper tubes is a small cost compared to the expense in time and effort of the rest of the study.

STABLE-ISOTOPE-LABELED INTERNAL STANDARD PEPTIDES

The use of stable-isotope-labeled internal standard (SIS) peptides allows accurate and reproducible quantitation, and the use of these internal standards will account for possible losses or suppression effects that occur "downstream" of the point where they are added. They cannot, however, compensate for effects that occur "upstream." In both SISCAPA,[7,8] immuno matrix-assisted laser desorption/ionization (iMALDI),[9,10] and our standard non-affinity-based assays of plasma workflows, SIS peptides are added after the digestion step—before the affinity enrichment in SISCAPA and iMALDI, and before the liquid chromatography-multiple reaction monitoring-mass spectrometry (LC-MRM-MS) analysis in our methods. This means that the use of these standard peptides will not compensate for non-reproducible or inefficient digestion.

Another consideration is that, because the final reported concentrations are all based on "known" concentrations of SIS peptides, these concentrations must be accurately determined. This is not trivial, especially at low levels. In our laboratory, SIS peptides are purified by HPLC, the peptide composition is determined by amino acid analysis, and the final purity is checked by capillary electrophoresis.

The ideal SIS peptide will incorporate amino acids isotopically labeled with ^{13}C or ^{15}N because these isotopes do not have an effect on the retention time of the peptide. The ideal mass difference between the endogenous and the labeled peptide is greater than 6 Da, to avoid overlapping isotope clusters, which would complicate the quantitation. The incorporation of deuterium as an isotopic label also causes a mass shift compared to the protonated form of the peptide, but deuterated peptides do not exactly co-elute with their unlabeled counterparts, and this complicates the development of the assay, as well as the quantitation step.

Because of the expense of synthesizing SIS peptides, this synthesis is usually one of the last steps in assay development—most of the assay is first done based on the endogenous peptides (in addition to Basic Local Alignment Search Tool [BLAST] searches to determine their uniqueness to the target protein). Peptides that have cysteines are usually avoided, not only because they can they form disulfide bonds, but because the possibility of oxidation may lead to multiple forms of the same peptide, thus complicating the quantitation (and reducing the sensitivity, if the signal from one peptide is spread out over multiple isoforms). Methionine and tryptophan are also avoided because of the potential for oxidation. The peptide should be within the most sensitive range of the mass spectrometer, for both detection of the precursor peptide and within an appropriate mass range for production of sequence-specific MS/MS fragment ions (this usually means below 2500 Da). There are several additional "rules" for target peptide detection, and these have been discussed elsewhere.[11–13]

There are software programs available that predict the MRM parameters. For example, both the open-source Skyline software[14] and the Applied Biosystems' MIDAS Workflow Designer[15] predict the optimal peptide charge-state based on the number of basic residues in the peptide, and also predict y-ion fragments with m/z values greater than the precursor ion m/z as suggested Q3 masses for the peptide MRM ion pairs. While these programs are good starting points, we have found that empirically tuning the mass spectrometer (charge state, declustering potential, and collision energy) gives the highest signal for a precursor-product ion pair and increases the sensitivity of the assay by a factor greater than 10 compared to using the suggested values.[12,13]

Serum Albumin	*Gelsolin, isoform 1*
Transferrin	*Ceruloplasmin*
Apolipoprotein A–I	**Complement Component C9**
Alpha–1–acid Glycoprotein 1	*Clusterin*
Apolipoprotein A–II Precursor	**Zinc–alpha–2–glycoprotein**
Haptoglobin Beta Chain	**Serum amyloid P–component**
Transthyretin	*Heparin Cofactor II*
Hemopexin	*Prothrombin*
Fibrinogen Gamma Chain	**Plasma Retinol–binding Protein Precursor**
Fibrinogen Alpha Chain	*Complement C4 Beta Chain*
Fibrinogen Beta Chain	*Complement C4 Gamma Chain*
Apolipoprotein C–III	**Apolipoprotein B–100**
Alpha–2–macroglobulin	*Alpha–2–antiplasmin*
Apolipoprotein C–I Lipoprotein	**Kininogen–1**
Alpha–1–antichymotrypsin	*Plasminogen*
Complement C3	*Apolipoprotein E*
Vitronectin	*Coagulation Factor XIIa Heavy Chain*
Inter–alpha–trypsin Inhibitor Heavy Chain H1	**Afamin**
Vitamin D–binding Protein	*Beta–2–glycoprotein I*
Alpha–1B–glycoprotein	*Angiotensinogen*
Antithrombin–III	*L–selectin*
Apolipoprotein A–IV	
Complement Factor H	
Complement B Factor	

Fig. 1. CVD biomarkers (italics) in the top 45 proteins in plasma.

DIGESTION

Digestion of a protein into peptides depends on the protein, the enzyme, and the digestion conditions. In proteomics, "specific" enzymes are usually used, meaning that the enzyme cleaves the protein only at specific, known amino acids or amino-acid sequences. This specificity reduces the complexity of the peptide identification step, and simplifies the prediction of the protein fragments for database searching. The most commonly used enzyme for protein digestion is trypsin, partly because it is inexpensive, and also because it cleaves after lysine and arginine residues,

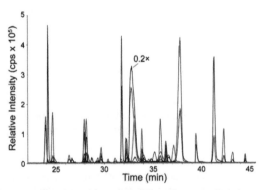

Fig. 2. RIC trace for the MRM-based assay for the 45 most abundant proteins in human plasma. RIC, reconstructed ion chromatogram. (*From* Kuzyk MA, Smith D, Yang J, et al. MRM-based, Multiplexed, Absolute Quantitation of 45 proteins in human plasma. Mol Cell Proteomics 2009;8:1860–77; with permission.)

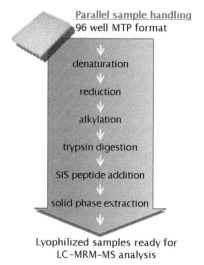

Parallel sample handling
96 well MTP format
⬇
denaturation
⬇
reduction
⬇
alkylation
⬇
trypsin digestion
⬇
SIS peptide addition
⬇
solid phase extraction
⬇
Lyophilized samples ready for
LC–MRM–MS analysis

Automated sample preparation
•liquid handling robotics
•increased throughput
•increased reproducibility
•decreased systematic bias

Fig. 3. Sample preparation protocol and automation.

resulting in a "reasonable" number of peptides per protein, and usually resulting in peptides of an appropriate size to be analyzed by MS. Moreover, the presence of the amino-terminus at one end and a free-amine-containing amino acid at the other end results in a peptide that usually produces good MS/MS fragmentation patterns, with b ions containing the N-terminus, and y ions containing the C-terminus.[16]

If you were designing an "ideal" digestion, you would denature the protein targets so that an enzyme could have access to all of the theoretical cleavage sites. Often, the first step is reduction and alkylation, designed to break and stabilize any disulfide (cys-cys) bonds. Next, you would completely unfold the protein. Commonly, this is done with detergents (eg, sodium dodecyl sulfate [SDS], deoxycholate, 3-[(3-chol-amidopropyl)dimethylammonio]-1-propanesulfonate) [CHAPS]), or solvents such as

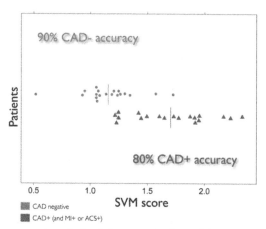

Fig. 4. Principle component analysis of the quantitation of these 5 proteins shows a clear separation between patients with CAD and those without CAD. CAD, coronary artery disease.

acetonitrile or methanol. However, enzymes are proteins, too. Fortunately, enzymes are naturally designed to be resistant to denaturation, but you must still be careful not to use conditions that are too harsh. If you have a specific target protein, the digestion should be optimized for the protein of interest. For the analysis of multiple targets (as in the assays described here), you must choose the best digestion conditions for reproducible and efficient digestion of the largest number of proteins "on average". In our studies, we have found deoxycholate to be the denaturant of choice.

"TOP-45" STUDY

Our first project for plasma analysis without depletion involved the direct MRM-based quantitation of the 45 most abundant proteins in plasma.[12] Twenty-eight of these proteins were previously identified as potential cardiovascular disease biomarkers (**Fig. 1**).[17] This assay was developed for use with an online HPLC system operating at a flow rate of 300 nL/min, and an AB 4000 QTRAP mass spectrometer. Three MRM transitions from one proteotypic peptide from each of the 45 proteins could be quantitated from undepleted human plasma in a single multiplexed 45-minute MRM assay consisting of 270 scheduled ion pairs.

- Adiponectin
- Afamin
- Albumin, serum
- Aldolase C
- Alpha-1-acid glycoprotein 1
- Alpha-1-antichymotrypsin
- Alpha-1-Anti-trypsin
- Alpha-1B-glycoprotein
- Alpha-2-antiplasmin
- Alpha-2-HS-glycoprotein
- alpha-2-macroglobulin
- Angiotensin-converting enzyme (ACE)
- Angiotensinogen
- Antithrombin-III
- Anti-trypsin
- Apolipoprotein A-I
- Apolipoprotein A-II precursor
- Apolipoprotein A-IV
- Apolipoprotein B-100
- Apolipoprotein C-I lipoprotein
- Apolipoprotein C-II
- Apolipoprotein C-III
- Apolipoprotein D
- Apolipoprotein E
- Apolipoprotein L1
- Apolipoprotein(a)
- Aspartate aminotransferase,mitochondrial (m-type)
- Beta-2-glycoprotein I
- CD105 (endoglin)
- Ceruloplasmin
- chitotriosidase
- Cholesterol ester transfer protein
- Chromogranin A
- Clusterin
- Coagulation Factor IX
- Coagulation Factor V
- Coagulation Factor VII
- Coagulation Factor VII-activating protease
- Coagulation Factor VIII

- Coagulation Factor X
- Coagulation Factor XI
- Coagulation factor XIIa heavy chain
- Coagulation Factor XIII A chain
- Coagulation Factor XIII B chain
- Complement C1 inactivator
- Complement C3
- Complement C4 beta chain
- Complement C4 gamma chain
- Complement component C9
- Complement factor B
- Complement factor H
- C-reactive protein (CRP)
- Creatine kinase-B
- Creatine kinase-MB M chain
- Endothelial cell protein C receptor
- Ferritin, Light chain
- Fibrinogen alpha chain
- Fibrinogen beta chain
- Fibrinogen gamma chain
- Fibrinopeptide A
- Fibronectin
- Follistatin
- Gelsolin, isoform 1
- Glial fibrillary acidic protein (GFAP)
- GPIIb/IIIa, soluble
- Haptoglobin beta chain
- Hemopexin
- Heparin cofactor II
- Histidine-rich glycoprotein
- IGF-1
- Insulin
- Insulin-like growth factor binding protein-1 (IGFBP-1)
- Inter-alpha-trypsin inhibitor HC1
- Intercellular adhesion molecule 1, soluble (sICAM-1)
- Interleukin-1 receptor family member, ST2
- Kininogen-1
- Leptin
- Leptin Receptor, Soluble

- L-selectin
- Macrophage colony-stimulating factor (MCSF)
- Matrix metalloproteinase-2 (MMP-2)
- Matrix metalloproteinase-9 (MMP-9)
- Myelin Basic Protein
- Myeloperoxidase
- Myoglobin, cardiac (Mb)
- Myosin heavy chain, cardiac Beta isoform (7)
- Myosin light chain I, cardiac
- natriuretic peptide, atrial, C-terminal (C-ANP)
- Neutrophil Protease-4
- NGAL
- Paraoxonase (PON 1)
- Paraoxonase (PON 3)
- Plasma retinol-binding protein precursor
- Plasminogen
- Plasminogen activator inhibitor (PAI)-1-antigen
- Platelet-activating factor (PAF) acetylhydrolase
- Platelet-derived growth factor B chain (PDGF)
- Prorenin
- Protein C
- Protein C inhibitor
- Protein S
- Protein Z
- Prothrombin
- Serum amyloid P-component
- Sex hormone-binding globulin
- Thrombospondin-1
- Tissue factor pathway inhibitor (TFPI)
- Tissue inhibitor of metalloproteinases-1 (TIMP-1)
- Tissue inhibitor of metalloproteinases-2 (TIMP-2)
- Transferrin
- Transforming growth factor-beta (TGF-beta)

- Transthyretin
- Tropomyosin 1 alpha chain
- Vitamin D-binding protein
- Vitronectin
- Von Willebrand Factor
- Zinc-alpha-2-glycoprotein

Fig. 5. Current MRM assay for 117 proteins in undepleted plasma (proteins marked in italics are known CVD biomarkers).

Fig. 2 shows the chromatogram resulting from the application of this assay to 1 μg of digested human plasma (14 nL of plasma) per analysis. The quantitation metrics for this assay were: r greater than 0.99 for 42 out of 45 assays, a lower limit of quantitation (LLOQ) in the attomole range for 27 of the 45 proteins, an analytical precision of less than 6% for 37 of the 45 proteins (replicate injections of the same sample into the LC-MRM-MS system), and a technical precision of less than 20% for 43 of the 45 proteins (3 analytical runs on separate days, including digestion and analysis).

A simple and easy-to-automate sample preparation procedure was developed and used for this assay (**Fig. 3**). The use of robotics for sample preparation increases the throughput of the system, decreases sample handling errors, and increases reproducibility.

USE OF THESE MULTIPLEXED ASSAYS FOR BIOMARKER DISCOVERY

For a clinical assay, the absolute quantities of the analytes in plasma must be known in order to determine if the parameters fall within the "normal" or "reference" range. This "top 45" assay was designed with this in mind. However, this assay can also be used as a biomarker discovery tool. Using this "top 45" MRM-based assay, we were able to find a panel of 5 protein biomarkers that correlated with acute coronary artery disease (CAD).[18] The separation between the patient groups is shown in **Fig. 4**. The

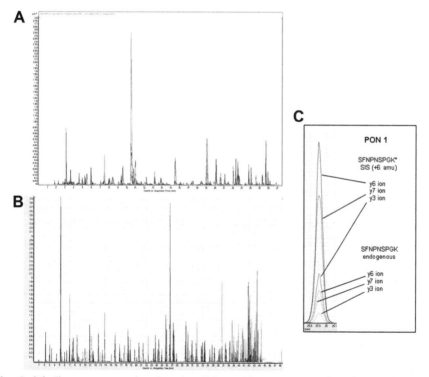

Fig. 6. (A) Chromatogram showing the 121-peptide-target MRM-based assay for CVD-related proteins in human plasma. (B) Chromatogram for 168 plasma peptides. (C) RIC traces for the MRM transitions for the SIS and endogenous forms of a proteotypic peptide.

panel of 5 proteins was 90% accurate in predicting the absence of CAD, and 80% accurate in predicting the presence of CAD in a group of 20 CAD-positive patients and 20 controls. Two of these 5 proteins were not previously known to be correlated with CAD.

Our current MRM assay was extended to include proteotypic peptides from the top 117 most abundant proteins in human plasma, 84 of which are known cardiovascular disease (CVD) biomarkers **(Fig. 5)**. Again, this assay is done without depletion and without enrichment. The chromatogram showing RIC traces for 726 MRM ion pairs is shown in **Fig. 6A**. This assay has since been expanded to include 168 proteotypic peptides **(Fig. 6B)**. An example of the MRM traces for the precursor and product ions for the SIS and endogenous forms of a proteotypic peptide is shown in **Fig. 6C**.

Several changes to the method had to be made to achieve this level of multiplexing. First, UPLC was used instead of nanoscale capillary LC. The UPLC gave improved retention time reproducibility, narrow peak widths, and it allowed more peaks to be detected and resolved in the same amount of chromatographic space.

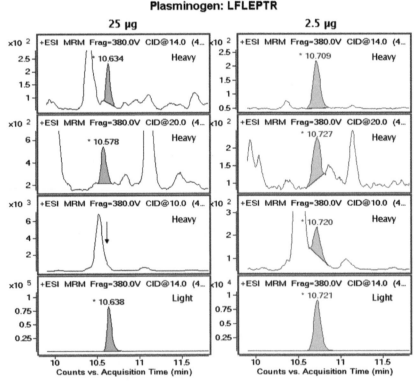

Fig. 7. Sensitivity increases with injection of less material because of improved S/N ratio. RIC traces shown in green are using the 10x diluted sample. The upper 3 are the RIC traces for the 3 transitions of the SIS versions of the plasminogen peptide LFLEPTR. The left-hand column shows the RICs obtained when 25 ug of total plasma protein was loaded on the column. The right-hand column shows the RICs obtained by analyzing the same sample, but 10 times diluted. Note the SIS/total protein ratio is the same in the samples, but the S/N ratios are higher in the diluted sample.

Protein	AB 4000 Qtrap	Agilent 6490 High Flow
PGC 103	Y	Y
PGC 121	Y	Y
PGC 17	Y	Y
PGC 18	Y	Y
PGC 23	Y	Y
PGC 26	Y	Y
PGC 64	Y	Y
PGC 97	Y	Y
PGC 100	Y	Y
PGC 29	Y	Y
PGC 137	Y	Y
PGC 222	Y	Y
PGC 230	Y	Y
PGC 233	Y	Y
PGC 221	Y	Y
PGC 69	Y	Y
PGC 77	Y	Y
PGC 108	Y	Y
PGC 125	N	Y
PGC 38	N	Y
PGC 104	N	Y
PGC 111	N	N
PGC 116	N	N
PGC 135	N	Y
PGC 223	N	Y
PGC 224	N	N

Fig. 8. Comparison of proteins detected in low-flow and high-flow systems.

The Agilent 1290 Infinity UPLC system was interfaced to a high-flow thermal gradient-focusing electrospray source, which uses heated sheath gas and an ion funnel, on an Agilent 6490 triple-quadrupole mass spectrometer. The use of higher flow rates allows for a faster assay, increasing the overall throughput. In addition, it facilitates the injection of larger sample volumes of a dilute sample without preconcentration. We have found that the use of higher flow rates and larger inner diameter (id) columns also allows the loading of larger sample amounts without overloading the column. The increased sample size compensates for any loss of sensitivity that might occur when going from nanoscale capillary LC (75 micron id; 300 nL/min) to minibore columns (2.1 mm id; 400 μL/min).

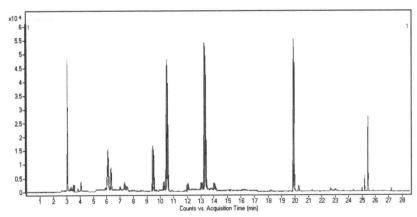

Fig. 9. Overlays of the chromatograms from runs 1, 10, 20, 30, 40, 50, 60, 70, 80, and 90.

METHOD OPTIMIZATION—AMOUNT INJECTED

Injections of 25 μg of a 10-fold diluted plasma digest (the equivalent of 2.5 μg of plasma injected) produced higher S/N values than injections of 25 μg of digested human plasma (**Fig. 7**). This is because of the reduced suppression effects in nanospray due to injecting less background.[19–21] Using this diluted sample resulted in an improved S/N ratio and we were able to detect 5 additional proteins that were below the detection limit on our original AB Sciex 4000 QTRAP system (**Fig. 8**).

RETENTION TIME STABILITY

The increased amount of column packing also results in a more robust system, allowing more injections without degradation of the system performance. **Fig. 9** shows an overlay of every tenth analysis (from 1 to 90) of the same nondepleted plasma sample on the same column, and demonstrates the stability and robustness of this newly developed high-flow system. For this figure, 12 SIS peptides were spiked into the sample at expected endogenous level. Four of these 12 peptides were not detectable with the older nano-flow QTRAP system.

Another visual display of the precision of this assay is shown graphically in **Fig. 10**A, which shows the retention time stability for 110 injections of the same sample (10 fmol SIS peptides [in 2.5 μg plasma digest], injected on-column). **Fig. 10**B shows the reproducibility metrics for 110 injections of the same sample using this high-flow method. All proteins showed a reproducibility in peak area of less than 10% RSD, and a less than 0.2% RSD in retention time. **Fig. 10**C shows the distribution of retention time % CV values for the analysis of 90 different patient plasma samples for 241 peptide targets. Statistically, most of the peptides (60%) gave a retention time precision of better than 0.25%, with 97% falling within 0.5% of the average value. This demonstrates the value of the retention time as an indicator that the correct peptide is being monitored, providing another level of confidence in the identification, in addition to the precursor and product ion masses.

QUANTITATION, DYNAMIC RANGE, AND PRECISION

A calibration curve for 6 ratios of plasminogen endogenous peptide to plasminogen SIS peptide, which was spiked into the solution to give relative ratios of 5 to 10,000,

A

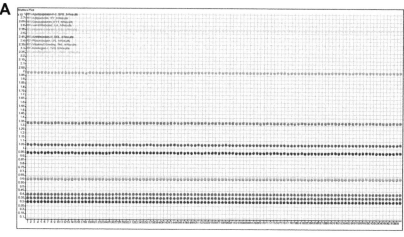

B

Protein	Response %RSD	Ret. Time %RSD
Adiponectin: IFYNQQNHYDGSTGK	9.8	0.13
Antithrombin–III : DDLYVSDAFHK	4.7	0.16
Apolipoprotein A–II precursor: SPELQAEAK	6.7	0.12
Apolipoprotein C–III: GWVTDGFSSLK	2.3	0.08
Ceruloplasmin : EYTDASFTNR	9.6	0.14
Heparin cofactor II: TLEAQLTPR	6.1	0.15
Histidine–rich glycoprotein: DGYLFQLLR	3.4	0.02
Kininogen–1 : TVGSDTFYSFK	3.3	0.13
L–selectin: AEIEYLEK	9.5	0.15
Plasminogen: LFLEPTR	2.2	0.13
Vitamin D–binding protein: THLPEVFLSK	3.0	0.12
von Willebrand Factor: ILAGPAGDSNVVK	9.5	0.15

C

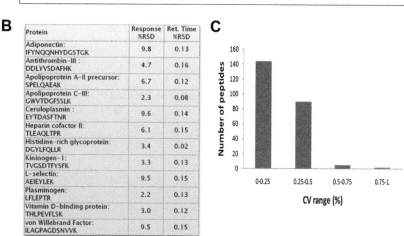

Fig. 10. (A) Retention time stability for 110 injections of 10 fmol of SIS peptides and 2.5 μg of plasma digest on-column (from top to bottom: Apolipoprotein A-II, Adipnectin, Ceruloplasmin, von Willebrand factor, Heparin cofactor II, L-selectin, antithrombin III, Plasminogen, Vitamin D-binding protein, Kininogen-1, Apolipoprotein C-III, and Histidine-rich glycoprotein). (B) Reproducibility metrics for 10 fmol SIS peptide spiked into 2.5 μg of digested plasma. (C) Retention time stability and precision for 90 patient plasma samples, analysed for 241 target peptides. Almost all of the peptide retention times were within the 0% to 0.05% CV range.

is shown in **Fig. 10**. Here, varying amounts of the plasminogen SIS peptide were spiked into 250 ng/μL nondepleted plasma digest. Three measurements were taken at each ratio of SIS/endogenous peptide. Excellent linearity and precision were observed over the concentration range from 5 to 10,000 amol/μL, giving an R^2 value of greater than .99 for a dynamic range of more than 3 orders of magnitude, with a precision RSD of 2.3%. (**Fig. 11**).

IMPLICATIONS FOR DIAGNOSTICS

MRM-based assays for human plasma proteins, with quantitation based on SIS peptides, have the specificity needed for clinical assays. This specificity is based on

A   Linear plot

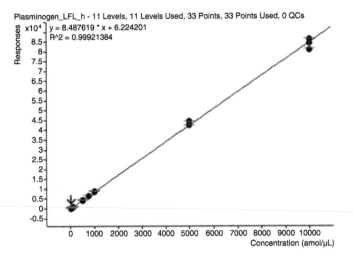

B   Log-Log plot

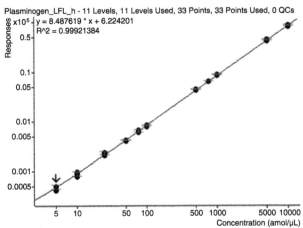

C

441.3>621.4	SIS peptide added (amol/µL)										
	5	10	25	50	75	100	500	750	1000	5000	10000
%Accuracy (average, n = 3)	98.2	97.5	104.6	98.1	99.2	97.4	101.3	101.8	101.3	101.8	98.8
Reproducibility (%RSD, n = 3)	11.85	10.78	7.21	1.17	6.22	2.74	1.41	0.48	1.52	2.58	3.24
Response factor	8.33	8.28	8.87	8.32	8.42	8.27	8.60	8.64	8.60	8.64	8.39
Precision (%RSD, n = 11)	2.30										

Fig. 11. Precision and reproducibility of high-flow LC-MRM-MS method on the Agilent 6490: Quantitation of the Plasminogen Peptide in Plasma. (*A*) Calibration curves for the observed ratio vs concentration. (*B*) Log-Log plot of the observed ratio vs concentration. (*C*) Accuracy, reproducibility, and precision of the assay over the range from 5 to 10,000 amol/µL SIS peptide added to 250 ng/µL plasma digest. All measurements were done in triplicate.

3 parameters of each proteotypic peptide: the mw, the requirement for co-elution with the SIS peptide, and the fragmentation of the precursor ion into a specific fragment ion. Our MRM-based assay uses the resolving power of UPLC and mass spectrometry to produce an assay for the 92 most abundant proteins in human plasma, in a simple automated process. We have demonstrated that this approach provides sufficient specificity so that affinity purification and depletion are not needed. Affinity purification, as used in SISCAPA and iMALDI, will still be needed for low-level analytes, but generating antibodies is expensive, and antibodies can have problems with cross-reactivity. For many analyses, affinity depletion or enrichment would only add additional and unnecessary steps to the protocol.

The approach described here has already been demonstrated for the 92 most abundant proteins in plasma, and we are continuing to add more proteins to this list. This accurate multiplexed MRM-based analysis can be done in approximately 1 hour, and is ready to be used for biomarker verification and validation studies requiring absolute quantitation of proteins, and panels of biomarker proteins, in human plasma.

ACKNOWLEDGMENTS

The University of Victoria-Genome BC Proteomics Centre is partially funded by platform grants from Genome-British Columbia and Genome Canada. We are also grateful for support from the NCE CECR Prevention of Organ Failure (PROOF) Centre of Excellence. Support for this research was also provided in part by the British Columbia Proteomics Network through a Small Projects Health Research Grant Award. The instrument used for the high-flow rate method was purchased in 2010 with funds obtained through a Strategic Opportunities Fund grant from Genome British Columbia and an award from Western Economic Diversification Canada.

REFERENCES

1. Anderson NL, Anderson NG. The human plasma proteome:history, character, and diagnostic prospects. Mol Cell Proteomics 2002;1:845–67.
2. Anderson L. Candidate-based proteomics in the search for biomarkers of cardiovascular disease. J Physiol 2005;563:23–60.
3. Proc JL, Kuzyk MA, Hardie DB, et al. A quantitative study of the effects of chaotropic agents, surfactants, and solvents on the digestion efficiency of human plasma proteins by trypsin. J Proteome Res 2010;9:5422–37.
4. Hulmes JD, Bethea D, Ho K, et al. An investigation of plasma collection, stabilization, and storage procedures for proteomic analysis of clinical samples. Clinical Proteomics 2004;1:17–31.
5. Aguilar-Mahecha A, Buchanan M, Kuzyk MA, et al. Comparison of blood collection tubes and processing protocols for plasma proteomics studies. Adriana Aguilar-Mahecha, Marguerite Buchanan, Michael A Kuzyk, Christoph Borchers, Mark Basik. Manuscript in preparation 2010.
6. Omenn GS, States DJ, Adamski M, et al. Overview of the HUPO Plasma Proteome Project: results from the pilot phase with 35 collaborating laboratories and multiple analytical groups, generating a core dataset of 3020 proteins and a publicly-available database. Proteomics 2005;5:3226–45.
7. Anderson NL, Anderson NG, Haines LR, et al. Mass spectrometric quantitation of peptides and proteins using stable isotope standards and capture by anti-peptide antibodies (SISCAPA), J Proteome Res 2004;3:235–44.
8. Anderson NL. High sensitivity quantitation of peptides by mass spectrometry. 2004; WO 2003US31126 20031002. 48 pp.

9. Borchers CH. Methods of quantitation and identification of peptides and proteins. U.S. Patent Office filed 12/02/2002, awarded 7/12/2010, US 7,846,748 B2.
10. Borchers CH. Methods of quantitation and identification of peptides and proteins. Canadian Patent Office filed 2002/12/02, awarded 2011/01/26, 2,507,864.
11. Kuzyk MA, Ohlund LB, Elliott MH. A comparison of MS/MS-based, stable-isotope-labeled, quantitation performance on ESI-quadrupole TOF and MALDI-TOF/TOF mass spectrometers. Proteomics 2009;9:3328–40.
12. Kuzyk MA, Smith D, Yang J, et al. MRM-based, Multiplexed, Absolute Quantitation of 45 proteins in human plasma. Mol Cell Proteomics 2009;8:1860–77.
13. Kuzyk MA, Parker CE, et al, editors. Methods in Molecular Biology; Humana Press; 2011.
14. Skyline_SRM/MRM_Builder, 2011, update v0.7, *Skyline Targeted Proteomics Environment*. Avaliable at: https://brendanx-uw1.gs.washington.edu/labkey/wiki/home/software/Skyline/page.view?name=default. Accessed July 9, 2011.
15. MIDAS_workflow_designer. Avaliable at: https://products.appliedbiosystems.com/ab/en/US/adirect/ab?cmd=catNavigate2&catID=604874&tab=DetailInfo. Accessed July 9, 2011.
16. Roepstorff P, Fohlman J. Proposal for a common nomenclature for sequence ions in mass spectra of peptides. Biomed Mass Spectrom 1984;11: 601.
17. Anderson L, Hunter CL. Quantitative mass spectrometric multiple reaction monitoring assays for major plasma proteins. Mol Cell Proteomics 2006;5: 573–88.
18. Cohen Freue GV, Borchers CH. Multiple reaction monitoring (MRM) – principles and application to coronary artery disease. Circ Cardiovasc Genet 2011. [Epub ahead of print].
19. Gangl ET, Annan M, Spooner N, et al. Reduction of signal suppression effects in ESI-MS using a nanosplitting device. Anal Chem 2001;73:5635–44.
20. Juraschek R, Dülcks T, Karas M. Nanoelectrospray—more than just a minimized-flow electrospray ionization source. J Am Soc Mass Spectrom 1999;10:300–8.
21. Fligge TA, Bruns K, Przybylski M. Analytical development of electrospray and nano-electrospray mass spectrometry in combination with liquid chromatography for the characterization of proteins. J Chromatoqr B Biomed Sci Appl 1998;706:91–100.

Peptide Immunoaffinity Enrichment Coupled with Mass Spectrometry for Peptide and Protein Quantification

Jeffrey R. Whiteaker, PhD, Amanda G. Paulovich, MD, PhD

KEYWORDS
- Quantitative proteomics • Selected reaction monitoring
- Stable isotope dilution • SISCAPA • Immuno-SRM
- Immuno-mass spectrometry

Quantitative protein assays are a critical component in exploring the relationship between protein abundance and biological features such as phenotype, disease, or response to treatment. Traditionally, the sandwich immunoassay (eg, enzyme-linked immunosorbent assay [ELISA]) has been the gold standard for protein quantification, owing to the high sensitivity, high throughput, and low per-sample cost (once an assay is developed and validated). However, the need for quantitative assays has far outpaced the rate of development. Modern strategies are capable of discovering hundreds to thousands of interesting genes and proteins that must be validated in follow-up experiments with a quantitative assay. Highly specific and sensitive assays are not available for quantifying the vast majority of human proteins, and *de novo* assay generation is associated with a high cost and long lead time. In addition, many existing assays suffer from poor specificity, a variety of interferences, lack of standardization, or all of the above. Thus, the current situation is severely limiting to all areas of research, from basic biology to clinical chemistry. We desperately need alternative strategies for building assays to any protein (or modification) of interest without a prohibitive investment in time, money, and other resources.

One technology that holds particular promise in improving the situation is quantitative targeted mass spectrometry. Selected reaction monitoring (SRM) is a targeted mass spectrometry technique that has increased sensitivity, compared to profiling modes of analysis, while maintaining high specificity for the target analyte. This method is well established in clinical reference laboratories for accurate quantification of small molecules in plasma, such as metabolites that accumulate as a result of

This work was funded by the National Cancer Institutes Clinical Proteomic Technology Assessment for Cancer (CPTAC) Program.
The author has nothing to disclose.
Fred Hutchinson Cancer Research Center, 1100 Fairview Avenue North, Seattle, WA 98109, USA
E-mail address: jwhiteak@fhcrc.org

Clin Lab Med 31 (2011) 385–396
doi:10.1016/j.cll.2011.07.004
0272-2712/11/$ – see front matter © 2011 Elsevier Inc. All rights reserved.

labmed.theclinics.com

inborn errors of metabolism.[1,2] Selected reaction monitoring has been utilized with increasing frequency in proteomics to measure the concentrations of target proteins in biological matrices.[3-8] To achieve quantitation of proteins, biological molecules are digested to component peptides using a proteolytic enzyme such as trypsin. One or more selected peptides whose sequences are unique to the target protein and are efficiently observed by the mass spectrometer (ie, "proteotypic" peptides) are then measured as quantitative stoichiometric surrogates for the protein of interest. Quantitation is performed by measuring the surrogate peptide relative to a spiked, stable isotope-labeled standard, using conventional stable isotope dilution methods.[9,10] The assays are specific, precise (CV ≤20%), multiplexable, and portable across laboratories and instrument platforms.[11,12] They are also relatively inexpensive to develop, especially compared to other quantitative technologies.

Currently, a serious limitation to more widespread use of SRM-based assays is the limited sensitivity typically achieved in complex samples. For example, without enrichment, SRM is typically able to measure proteins present in the 100 to 1000 ng/mL concentration range in plasma,[7] although many "biologically interesting" proteins are found several orders of magnitude below that range. An enrichment or fractionation step can enhance the sensitivity and extend detection to low abundance analytes. For example, previous studies have demonstrated the success of using abundant protein depletion with limited strong cation exchange fractionation[13] or glycopeptide enrichment[14] to analyze proteins in the low ng/mL range. However, this is unattractive for analysis of large numbers of samples because of the increase in cost and time associated with extra sample handling, as well as the potential impact of multiple sample handling steps on measurement variability and analyte recovery.

Another approach for improving sensitivity is to employ immunoaffinity techniques for selective enrichment of the analytes. Once captured, the enriched analyte is quantified using mass spectrometry. Several modes of implementation are possible[15-18] using antibodies for proteins or peptides. The design of the assay depends on several considerations, including reagent availability, laboratory resources, requirements for throughput and sample handling, and the nature of the target (ie, targeting specific forms or modifications can dictate how the assay is configured). One approach is to use antipeptide antibodies to capture endogenous (ie, light) peptides and a stable isotope-labeled (ie, heavy) peptide internal standard (**Fig. 1**). This enrichment approach is referred to as Stable Isotope Standards with Capture by Antipeptide Antibodies (SISCAPA)[19] and, when coupled with quantification using SRM-targeted mass spectrometry, can generally be referred to as an immuno-SRM assay. This chapter will review recent advancements in the area of peptide immuno-SRM assay development, with a focus on technological aspects and recent applications.

ADVANTAGES OF IMMUNO-SRM ASSAYS

Coupling peptide immunoaffinity enrichment with mass spectrometry in an immuno-SRM assay has many advantages compared with traditional immunoassays and SRM assays lacking enrichment (**Table 1**). First, compared with traditional sandwich immunoassays, immuno-SRM requires a single antibody with relaxed specificity requirements. The mass spectrometer acts like a second antibody with very high specificity. Restrictions typical to antibody development in traditional sandwich immunoassays are avoided because there are no requirements for a specific epitope and no constraints for finding multiple antibodies with independent epitope recognition. This substantially decreases the time involved in screening for working antibodies. Using a single antibody also simplifies the assay design and ultimately reduces the

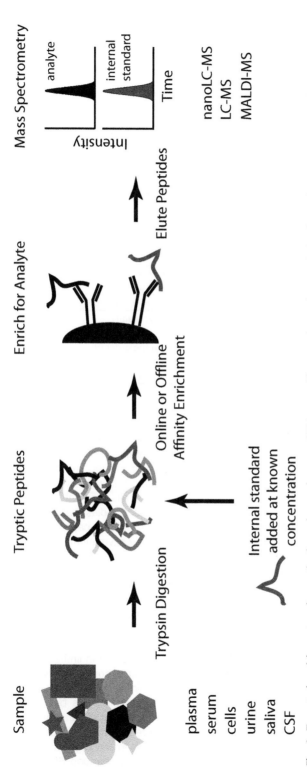

Fig. 1. Targeted enrichment and quantitation using an immuno-SRM assay. The sample can be a variety of complex proteomic samples. To achieve quantitation of the targeted protein(s), these larger molecules are digested to component peptides using an enzyme such as trypsin. A stable isotope standard (SIS) is added to the sample at a known concentration for quantitative analysis. The selected peptides are then enriched using antipeptide antibodies immobilized on a solid support. Following washing and elution from the antipeptide antibody, the amount of surrogate peptide is measured relative to the stable isotope standard using targeted mass spectrometry.

Table 1
Summary of advantages and performance features of immuno-SRM assays based on peptide enrichment

Attribute	Description	Example(s)
Time	Assays configured in less than a year, multiplexing allows development in parallel	[30]
Cost	Synthetic peptides and a single antibody per target	[30]
Standards	Quality-controlled and quantifiable reagents	[11]
High specificity of mass spectrometer	Relaxed requirements on antibody, reduces time to get working antibody	
Avoids autoantibody and antiregeant interference	Autoantibodies and interfering proteins are digested in sample preparation, eliminating interactions with analytes and reagents	[26]
Versatility in detecting protein modifications	Modified peptides can be enriched and targeted for quantification	[44,51]
Large dynamic range	Greater than 3 orders of magnitude	[27,50]
Multiplexing	Currently tens of assays readily multiplexed	[24]
Sensitivity	Limits of detection on the order of ng/mL or better	[23,26,27]
Precision	Less than 20% coefficient of variation	[24]

cost per sample. Another advantage of immuno-SRM assays is the standardization of synthetic peptide reagents. The use of stable isotope-labeled peptides as standards provides a reagent whose purity can be greatly controlled and highly characterized, the amount can be quantified using standard procedures (eg, amino acid analysis), and the standard can be used in many laboratories with concordant results. Furthermore, enrichment of peptides avoids many pitfalls associated with traditional immunoassays. Interferences by endogenous autoantibodies (recognizing the targeted protein) or antireagent antibodies (recognizing the capture antibody) are avoided because these endogenous proteins are enzymatically digested in the sample preparation.[20-22] Finally, the ability to readily multiplex a number of targets into a single assay is a tremendous advantage for immuno-SRM. Examples have been demonstrated for tens of peptides in a single assay[23,24] but, theoretically, much higher limits are possible.

The analytical performance of the assays is acceptable for a range of applications. Enrichment of the target peptide over 1000-fold is possible.[19-25] This approach also provides reduced ion suppression, enabling very sensitive detection of peptides enriched from complex samples. Demonstrations of detection limits in plasma have shown it is possible to analyze ng/mL protein concentrations with dynamic ranges over 3 orders of magnitude. Extending detection limits even lower is possible by increasing the amount of input sample.[23] The repeatability has been examined in several publications.[19,23,25-27] In most cases, the coefficient of variation for complete process replicates is below 20%. However, there are instances of individual proteins that show higher CVs, indicating some optimization in the digestion, and capture protocols may be necessary for specific analytes. Overall, immuno-SRM assays show good figures of merit and a number of advantages over immunoassays and SRM alone.

REAGENT AND ASSAY DEVELOPMENT

There are several steps involved in developing an immuno-SRM assay. These include target peptide selection, reagent generation, optimization, and configuration. Assay development begins with target peptide selection. Once a given protein or proteins are targeted, the general rules associated with picking peptides for SRM assays apply the same to immuno-SRM development.[9] These include selecting tryptic peptides that are unique to the protein of interest, that respond well in mass spectrometry (relatively good ionization and production of several fragments suitable for SRM development), and that do not contain known modifications (unless the modification is specifically being targeted). It is most common to select tryptic peptides in lengths of 8 to 22 amino acids with moderate hydrophobicity (very hydrophilic and very hydrophobic peptides are less stable due to retention time variation in high pressure liquid chromatography and loss to surfaces). Methionine residues (oxidation), N-terminal glutamine (cyclization), asparagine followed by glycine or proline (prone to deamidation), and dibasic termini (eg, neighboring lysine or arginine residues such as KK, KR, RR, RK have the potential for variable digestion efficiency) are undesirable. For quantitation, stable isotope-labeled peptides are used as internal standards. It is important to ensure the labeled peptide is sufficiently separated in mass from the analyte peptide, ideally containing a mass shift that can be measured in the precursor and fragment ions. Thus, it is typical to incorporate heavy (eg, ^{13}C and ^{15}N) labeled amino acids at the C-terminus of the peptide (ie, K- or R-labeled) and design SRM transitions targeting y-ions.

There are a variety of antigenicity prediction algorithms available when developing antibodies to protein targets. These predictors are based principally on the location of the peptide sequence within the 3-dimensional protein structure (ie, the extent of surface exposure or location in beta-turns).[28] However, these prediction algorithms have not yet found good utility in generating antibodies for the linear epitopes found in tryptic peptides used for immuno-SRM assays.[29,30] As more assays are evaluated, there may be an improvement in the development of prediction algorithms.

The success rate in generating affinity reagents for use in immuno-SRM is relatively high. A large scale evaluation of rabbit polyclonal antibodies showed good success (>50%) for making individual antipeptide antibodies with detection limits in the ng/mL-to-tens-of-ng/mL range.[30] By making antibodies to multiple peptides per protein, the success rate for generating an assay to a given protein of interest can be much higher (>90%). The antibody generation process can be multiplexed by injecting several peptides into a single animal, providing time and cost savings. The maximum number of multiplexed immunogens possible before the immune response to individual targets is diminished (ie, immunodominance) has not been characterized. The time required to generate and characterize polyclonal antibodies makes it feasible for a laboratory to develop hundreds of assays per year, a considerable advantage compared to traditional immunoassays. Finally, polyclonal antibodies have a typical yield sufficient to generate assays for hundreds to thousands of samples, suitable for large biological experiments or biomarker verification studies.

While affinity-purified polyclonal antibodies offer an attractive and affordable source of affinity reagents, monoclonal antibodies are far more advantageous for many applications. Monoclonal antibodies provide a renewal affinity reagent, providing the benefits of a standardized reagent and the capability of usage in an unlimited number of samples. Furthermore, monoclonals offer the potential to obtain higher binding affinities because of screening for the best-performing clonal hybridomas. The use of a monoclonal antibody in an immuno-SRM assay response curve has been demonstrated to have good performance,[31] and methods have been developed for determining the binding characteristics of antipeptide monoclonal antibodies using a refined surface plasmon resonance technique.[32] These studies show the capabilities and promise of using monoclonals. Methods to increase the throughput of screening for high affinity antibodies using protocols near the final assay format have also been implemented on both electrospray[31] and Matrix Assisted Laser Desorption Ionization (MALDI) platforms.[33] Extending the approach to higher throughput makes the large-scale production of monoclonal reagents a distinct possibility.

In addition to analyte-specific antibodies, group-specific antipeptide antibodies can achieve greater coverage of the proteome using a single affinity reagent. Triple X Proteomics (TXP) antibodies[34] are designed for linear epitopes of 3 to 4 amino acids and the terminally charged group of peptides. The short sequence comprising the epitope can be chosen to overlap among many proteins; thus, the antibodies recognizing that epitope will enrich a whole class of peptides. Initial calculations show TXP antibodies can cover a large fraction of the human proteome with a fraction of the antibodies required for peptide-specific assays.[35] Demonstration of the approach showed success in enriching 38 signature peptides from cell lysates using 2 TXP antibodies.[36] Issues regarding the performance of these antibodies (ie, analyte recovery, sensitivity, reproducibility) in comparison to other peptide-specific approaches remain to be characterized, but the approach offers promise for efficient assay development to large numbers of proteins.

Once affinity reagents and synthetic peptides are obtained, they are assembled into the working assay. Configurations for peptide immunoaffinity enrichment are presented below. The reader is referred to several excellent reviews for more information on SRM assay configuration and optimization.[9,10]

ASSAY CONFIGURATION

There are 2 predominant formats available for immunoaffinity enrichment of peptides. One format is enrichment using an affinity column, where antibodies are covalently coupled to chromatography media.[19,37-39] The column format offers the advantage of an online analysis using established affinity chromatography techniques. The use of POROS (Applied Biosystems, Foster City, CA, USA) nanocolumns was originally used in SISCAPA enrichment[19] and features high binding capacity, a relatively high concentration of antibodies allowing for rapid enrichment of target peptides, and the ability to prepare columns with a variety of functionalized groups. The column can also be regenerated and used to analyze many samples (eg, hundreds) depending on the antibody and assay parameters. These features make online immunoaffinity enrichment attractive from a ruggedness/robustness standpoint. Disadvantages include the potential for sample carryover, inflexibility in changing the analytes, and difficulties passing a large sample volume over the column.

An alternative to immunoaffinity chromatography is offline enrichment using magnetic particles.[25,27] Magnetic particles are available in a wide array of chemistries allowing for coupling to antibodies. Particles coated in Protein G are commonly employed, as they offer the binding of antibodies in a preferred orientation. Offline enrichment is advantageous in that samples can be processed in parallel using larger volumes of relatively viscous samples or samples that might otherwise be problematic in chromatography systems. Magnetic particle processing has been automated in 96 well plates for the SISCAPA enrichment step, with elution in the plates for analysis by mass spectrometry.[23] Alternatively, a novel bead trap device was developed to perform the bead-handling steps inline with the nanoflow chromatography system.[40] This minimizes losses of peptides to containers between the elution and analysis steps. Offline enrichment has also been implemented by immobilizing antibodies in pipet tips.[41]

In addition to electrospray mass spectrometry, MALDI has been used extensively in conjunction with peptide immunoaffinity enrichment. The largest advantage of MALDI, compared to electrospray, is the potential for detection of the enriched peptide without a chromatography step. This enables the analysis to be performed in a very short time (requiring minutes or less), enabling truly high throughput measurements. An additional advantage of MALDI is a greater tolerance to contaminating salts or detergents. The combination of immunoaffinity enrichment and MALDI detection was first configured with high throughput capabilities using the affinity reagent immobilized in pipet tips.[41] Although most commonly employed to analyze protein variants,[42,43] this approach is also capable of analyzing smaller peptide analytes.[44] A particular strength of this mass spectrometric immunoassay approach is the characterization of protein variants across a large number of samples. For targeting specific peptides using commercially available antibodies, assays have been configured with detection and quantification by MALDI covering 2 orders of magnitude.[45] Quantitative assays have also been configured for peptides from epidermal growth factor receptor[46], IgIC—a protein used for detection of *Francisella tularensis*,[47] and angiotensin.[48] The use of MALDI will likely continue to grow due to the tremendous potential for high throughput measurements.

APPLICATIONS

There are several notable examples of immuno-SRM assays employed in the literature. The most prevalent use of the assays has been in the area of preclinical verification of biomarkers. For example, an assay for Fibulin-2 was developed to verify its discovery as a novel circulating marker in a mouse model of breast cancer.[49] The assay had a dynamic range of over 2 orders of magnitude and was employed to verify the overabundance (30-fold) of Fibulin-2 in the plasma of mice with tumors. In this case, the assay was developed to increase the sensitivity for the target protein, which could not be quantitated by SRM alone. In another example, a multiplexed immuno-SRM assay was developed for Troponin I, an established marker of cardiac injury, and interleukin-33, an emerging cardiovascular biomarker.[27] The intent was to demonstrate the assay performance in a verification setting. The assays were easily multiplexed and showed sufficient precision, reproducibility, and sensitivity for utilization in verification studies. Finally, a highly multiplexed immuno-SRM assay was used to evaluate the potential for mass spectrometry-based technologies to impact a biomarker pipeline.[24] The 31-plex assay was used to provide initial biomarker verification in a mouse model of cancer by assaying 80 individual samples. The assay showed good reproducibility with a median limit of detection of 26 ng/mL (using 10 μL plasma).

The choice of surrogate peptide should be carefully considered and can provide additional possibilities in designing an immuno-SRM assay. An example of the versatility in building assays to surrogate peptides was illustrated in the development of an assay for pepsin and pepsinogen.[50] Peptide immunoaffinity enrichment was coupled with mass spectrometry for quantitating pepsin in saliva using an online column enrichment format. Notably for this assay, the C-terminal fragment peptide was chosen as the surrogate peptide to provide a combined measurement of pepsin and pepsinogen. The assay was capable of measuring pepsin concentrations in the low picomolar range with good reproducibility. Furthermore, the method was scalable to analyze larger volumes of initial material to customize the assay to expected sample concentrations. Another example highlighting the versatility in choosing peptides was the analysis of aberrant glycoforms of Tissue Inhibitor of Metalloproteinase 1 (TIMP1).[51] In this case, an immuno-SRM assay for TIMP1 was combined with lectin enrichment of the aberrantly glycosylated proteins in the serum of colon cancer patients. The assay was able to achieve detection limits below 1 ng/mL for the targeted isoforms using the combined enrichment approach.

Finally, there is the potential for immuno-SRM assays to be of great utility in the clinic. Thyroglobulin, a particularly challenging protein to assay,[52] was quantified by immuno-SRM to provide an assay free of the interferences affecting traditional immunoassays.[26] A limit of detection of 2.6 ng/mL was achieved using the polyclonal antibody generated against a target peptide for thyroglobulin. Comparison to existing immunoassays showed good correlation and agreement, suggesting the assay was nearly good enough for clinical use. Improvements in the overall sensitivity, through higher affinity monoclonal antibodies or advancements in new mass spectrometers, should make the assay attractive for patient sample analysis.

FUTURE AREAS OF DEVELOPMENT

As presented above, the performance, multiplex ability, ease of configuration, versatility, and specificity are considerable advantages to using immuno-SRM assays to meet the demand for quantitative protein assays. However, while the technology is promising, there remain several areas where the technique continues to see growth.

One area where optimization will be required is trypsin digestion. The imperfect nature of trypsin digestion, for which no current standards exist, is a source of error in using proteotypic peptides as stoichiometric surrogates of protein abundance. Incomplete tryptic digestion of parent proteins can result in underestimation of protein concentrations. One solution is to use stable isotope-labeled proteins as standards in immunoaffinity enrichment coupled to quantitative mass spectrometry.[53,54] A protein standard can be quantified independently (using amino acid analysis) and used as a reference for absolute protein quantification. This has the advantage of encompassing all analytical aspects of the assay, including trypsin digestion; this, however, is also imperfect since posttranslational modifications affecting trypsin digestion in the biospecimen under analysis may not be present in the recombinant protein standards. The reproducibility of digestion may be more critical than the completion of trypsin digestion since precise relative quantitation could be achieved if the digestion were incomplete, yet reproducible. The efficiency of trypsin digestion is likely to vary considerably among proteins; hence, for each analyte, it will be critical to take measures to minimize the effects of variable digestion and to detect it when it occurs.[55]

A major cost associated with the development of any immuno-SRM assay is the generation of antibodies. One approach for reducing the cost is to multiplex the immunization process.[30] This approach also has the benefit of increasing the success rate for a given protein target. Other strategies for affinity reagent development must be developed to significantly reduce the cost. It remains to be seen whether less expensive, recombinant approaches to antibody production will be successful in making high affinity antipeptide antibodies. In addition, the use of alternative reagents, like DNA aptamers,[56] may provide a versatile tool for peptide enrichment. Finally, the potential for multiplexing measurements using immuno-SRM assays has not been fully explored. We recently investigated the performance of multiplexing 50 peptides together into one assay[57] with good success. However, the mass spectrometer is capable of much higher multiplexing without compromising specificity.

Thus, future work will likely include the analysis of hundreds of analytes in a single assay, a tremendous advantage for saving the volume of precious samples and reducing the cost per analyte.

SUMMARY

The ability to readily develop highly multiplexed, sensitive assays to any group of proteins would transform biological science. Such assays would enable large-scale investigations into fundamental biology, improve our understanding of the relationships between molecular biology and disease, and provide a means to narrow the gap between biomarker discovery and personalized clinical medicine. Technologies utilized in such an endeavor must be versatile, time- and cost-effective. Immunoaffinity enrichment coupled with mass spectrometry has shown tremendous potential for meeting the demands of quantitative protein assays. The immuno-SRM approach provides a cost-effective alternative to traditional immunoassays, with good performance capabilities and tremendous versatility.

REFERENCES

1. Want EJ, Cravatt BF, Siuzdak G. The expanding role of mass spectrometry in metabolite profiling and characterization. Chembiochem 2005;6:1941–51.
2. Chace DH, Kalas TA. A biochemical perspective on the use of tandem mass spectrometry for newborn screening and clinical testing. Clin Biochem 2005;38:296–309.
3. Barr JR, Maggio VL, Patterson DG Jr, et al. Isotope dilution–mass spectrometric quantification of specific proteins: model application with apolipoprotein A-I. Clin Chem 1996;42:1676–82.
4. Gerber SA, Rush J, Stemman O, et al. Absolute quantification of proteins and phosphoproteins from cell lysates by tandem MS. Proc Natl Acad Sci U S A 2003;100:6940–5.
5. Kuhn E, Wu J, Karl J, et al. Quantification of C-reactive protein in the serum of patients with rheumatoid arthritis using multiple reaction monitoring mass spectrometry and 13C-labeled peptide standards. Proteomics 2004;4:1175–86.
6. Barnidge DR, Goodmanson MK, Klee GG, et al. Absolute quantification of the model biomarker prostate-specific antigen in serum by LC-Ms/MS using protein cleavage and isotope dilution mass spectrometry. J Proteome Res 2004;3:644–52.
7. Anderson L, Hunter CL. Quantitative mass spectrometric multiple reaction monitoring assays for major plasma proteins. Mol Cell Proteomics 2006;5:573–88.
8. Agger SA, Marney LC, Hoofnagle AN. Simultaneous quantification of apolipoprotein A-I and apolipoprotein B by liquid chromatography–multiple reaction monitoring/mass spectrometry. Clin Chem 2010;56:1804–13.
9. Lange V, Picotti P, Domon B, et al. Selected reaction monitoring for quantitative proteomics: a tutorial. Mol Syst Biol 2008;4:222.
10. Pan S, Aebersold R, Chen R, et al. Mass spectrometry based targeted protein quantification: methods and applications. J Proteome Res 2009;8:787–97.
11. Addona TA, Abbatiello SE, Schilling B, et al. Multi-site assessment of the precision and reproducibility of multiple reaction monitoring-based measurements of proteins in plasma. Nat Biotechnol 2009;27:633–41.
12. Kuzyk MA, Smith D, Yang J, et al. MRM-based, multiplexed, absolute quantitation of 45 proteins in human plasma. Mol Cell Proteomics 2009;8:1860–77.
13. Keshishian H, Addona T, Burgess M, et al. Quantitative, multiplexed assays for low abundance proteins in plasma by targeted mass spectrometry and stable isotope dilution. Mol Cell Proteomics 2007;6:2212–29.

14. Zhou Y, Aebersold R, Zhang H. Isolation of N-linked glycopeptides from plasma. Anal Chem 2007;79:5826–37.
15. Ackermann BL, Berna MJ. Coupling immunoaffinity techniques with MS for quantitative analysis of low-abundance protein biomarkers. Expert Rev Proteomics 2007;4: 175–86.
16. Nedelkov D. Mass spectrometry-based immunoassays for the next phase of clinical applications. Expert Rev Proteomics 2006;3:631–40.
17. Kiernan UA. Quantitation of target proteins and post-translational modifications in affinity-based proteomics approaches. Expert Rev Proteomics 2007;4:421–8.
18. Parker CE, Pearson TW, Anderson NL, et al. Mass-spectrometry-based clinical proteomics–a review and prospective. Analyst 2010;135:1830–8.
19. Anderson NL, Anderson NG, Haines LR, et al. Mass spectrometric quantitation of peptides and proteins using Stable Isotope Standards and Capture by Antipeptide Antibodies (SISCAPA). J Proteome Res 2004;3:235–44.
20. Hoofnagle AN, Wener MH. The fundamental flaws of immunoassays and potential solutions using tandem mass spectrometry. J Immunol Methods 2009;347:3–11.
21. Tate J, Ward G. Interferences in immunoassay. Clin Biochem Rev 2004;25:105–20.
22. Ismail AA. Interference from endogenous antibodies in automated immunoassays: what laboratorians need to know. J Clin Pathol 2009;62:673–8.
23. Whiteaker JR, Zhao L, Anderson L, et al. An automated and multiplexed method for high throughput peptide immunoaffinity enrichment and multiple reaction monitoring mass spectrometry-based quantification of protein biomarkers. Mol Cell Proteomics 2010;9:184–96.
24. Whiteaker JR, Lin C, Kennedy J, et al. A targeted proteomics-based pipeline for verification of biomarkers in plasma. Nat Biotechnol 2011;29:625–34.
25. Whiteaker JR, Zhao L, Zhang HY, et al. Antibody-based enrichment of peptides on magnetic beads for mass-spectrometry-based quantification of serum biomarkers. Anal Biochem 2007;362:44–54.
26. Hoofnagle AN, Becker JO, Wener MH, et al. Quantification of thyroglobulin, a low-abundance serum protein, by immunoaffinity peptide enrichment and tandem mass spectrometry. Clin Chem 2008;54:1796–804.
27. Kuhn E, Addona T, Keshishian H, et al. Developing multiplexed assays for Troponin I and Interleukin-33 in plasma by peptide immunoaffinity enrichment and targeted mass spectrometry. Clin Chem 2009;55:1108–17.
28. Greenbaum JA, Andersen PH, Blythe M, et al. Towards a consensus on datasets and evaluation metrics for developing B-cell epitope prediction tools. J Mol Recognit 2007;20:75–82.
29. Blythe MJ, Flower DR. Benchmarking B cell epitope prediction: underperformance of existing methods. Protein Sci 2005;14:246–8.
30. Whiteaker JR, Zhao L, Abbatiello SE, et al. Evaluation of large scale quantitative proteomic assay development using peptide affinity-based mass spectrometry. Mol Cell Proteomics 2011;10:M110.005645.
31. Schoenherr RM, Zhao L, Whiteaker JR, et al. Automated screening of monoclonal antibodies for SISCAPA assays using a magnetic bead processor and liquid chromatography-selected reaction monitoring-mass spectrometry. J Immunol Methods 2010;353:49–61.
32. Pope ME, Soste MV, Eyford BA, et al. Antipeptide antibody screening: selection of high affinity monoclonal reagents by a refined surface plasmon resonance technique. J Immunol Methods 2009;341:86–96.

33. Razavi M, Pope ME, Soste MV, et al. MALDI immunoscreening (MiSCREEN): a method for selection of antipeptide monoclonal antibodies for use in immunoproteomics. J Immunol Methods 2011;364:50–64.
34. Poetz O, Hoeppe S, Templin MF, et al. Proteome wide screening using peptide affinity capture. Proteomics 2009;9:1518–23.
35. Planatscher H, Supper J, Poetz O, et al. Optimal selection of epitopes for TXP-immunoaffinity mass spectrometry. Algorithms Mol Biol 2010;5:28.
36. Hoeppe S, Schreiber TD, Planatscher H, et al. Targeting peptide termini, a novel immunoaffinity approach to reduce complexity in mass spectrometric protein identification. Mol Cell Proteomics 2011;10:M110.002857.
37. Berna M, Schmalz C, Duffin K, et al. Online immunoaffinity liquid chromatography/tandem mass spectrometry determination of a type II collagen peptide biomarker in rat urine: investigation of the impact of collision-induced dissociation fluctuation on peptide quantitation. Anal Biochem 2006;356:235–43.
38. Li WW, Nemirovskiy O, Fountain S, et al. Clinical validation of an immunoaffinity LC-MS/MS assay for the quantification of a collagen type II neoepitope peptide: a biomarker of matrix metalloproteinase activity and osteoarthritis in human urine. Anal Biochem 2007;369:41–53.
39. Nemirovskiy O, Li WW, Szekely-Klepser G. Design and validation of an immunoaffinity LC-MS/MS assay for the quantification of a collagen type II neoepitope peptide in human urine: application as a biomarker of osteoarthritis. Methods Mol Biol 2010; 641:253–70.
40. Anderson NL, Jackson A, Smith D, et al. SISCAPA peptide enrichment on magnetic beads using an in-line bead trap device. Mol Cell Proteomics 2009;8:995–1005.
41. Nelson RW, Krone JR, Bieber AL, et al. Mass spectrometric immunoassay. Anal Chem 1995;67:1153–8.
42. Oran PE, Jarvis JW, Borges CR, et al. Mass spectrometric immunoassay of intact insulin and related variants for population proteomics studies. Proteomics Clin Appl 2011. [Epub ahead of print].
43. Trenchevska O, Nedelkov D. Targeted quantitative mass spectrometric immunoassay for human protein variants. Proteome Sci 2011;9:19.
44. Oran PE, Jarvis JW, Borges CR, et al. C-peptide microheterogeneity in type 2 diabetes populations. Proteomics Clin Appl 2010;4:106–11.
45. Warren EN, Elms PJ, Parker CE, et al. Development of a protein chip: a MS-based method for quantitation of protein expression and modification levels using an immunoaffinity approach. Anal Chem 2004;76:4082–92.
46. Jiang J, Parker CE, Hoadley KA, et al. Development of an immuno tandem mass spectrometry (iMALDI) assay for EGFR diagnosis. Proteomics Clin Appl 2007;1: 1651–9.
47. Jiang J, Parker CE, Fuller JR, et al. An immunoaffinity tandem mass spectrometry (iMALDI) assay for detection of Francisella tularensis. Anal Chim Acta 2007;605:70–9.
48. Reid JD, Holmes DT, Mason DR, et al. Towards the development of an immuno MALDI (iMALDI) mass spectrometry assay for the diagnosis of hypertension. J Am Soc Mass Spectrom 2010;21:1680–6.
49. Whiteaker JR, Zhang H, Zhao L, et al. Integrated pipeline for mass spectrometry-based discovery and confirmation of biomarkers demonstrated in a mouse model of breast cancer. J Proteome Res 2007;6:3962–75.
50. Neubert H, Gale J, Muirhead D. Online high-flow peptide immunoaffinity enrichment and nanoflow LC-MS/MS: assay development for total salivary pepsin/pepsinogen. Clin Chem 2010;56:1413–23.

51. Ahn YH, Lee JY, Lee JY, et al. Quantitative analysis of an aberrant glycoform of TIMP1 from colon cancer serum by L-PHA-enrichment and SISCAPA with MRM mass spectrometry. J Proteome Res 2009;8:4216–24.

52. Hoofnagle AN, Wener MH. Serum thyroglobulin: a model of immunoassay imperfection. Clin Lab Int 2006;12:12–4.

53. Kippen AD, Cerini F, Vadas L, et al. Development of an isotope dilution assay for precise determination of insulin, C-peptide, and proinsulin levels in non-diabetic and type II diabetic individuals with comparison to immunoassay. J Biol Chem 1997;272: 12513–22.

54. Janecki DJ, Bemis KG, Tegeler TJ, et al. A multiple reaction monitoring method for absolute quantification of the human liver alcohol dehydrogenase ADH1C1 isoenzyme. Anal Biochem 2007;369:18–26.

55. Proc JL, Kuzyk MA, Hardie DB, et al. A quantitative study of the effects of chaotropic agents, surfactants, and solvents on the digestion efficiency of human plasma proteins by trypsin. J Proteome Res 2010;9:5422–37.

56. Zhao Y, Widen SG, Jamaluddin M, et al. Quantification of activated NF-{kappa}B/RelA complexes using ssDNA aptamer affinity - stable isotope dilution–selected reaction monitoring–mass spectrometry. Mol Cell Proteomics 2011;10:M111.008771.

57. Whiteaker JR, Zhao L, Lin C, et al. Sequential multiplexed analyte quantification using peptide immunoaffinity enrichment coupled to mass spectrometry. Mol Cell Proteomics 2011. [Epub ahead of print].

The Analysis of Native Proteins and Peptides in the Clinical Lab Using Mass Spectrometry

Cory E. Bystrom, PhD

KEYWORDS
- Mass spectrometry • Clinical diagnostics • Proteomics
- Peptide • Protein

Mass spectrometric analysis of proteins and peptides has now enjoyed more than 2 decades of vigorous development and growth. As a key analytical technology that has been used with great success for the analysis of small molecules, the development of mass spectrometry instrumentation, software, and informatics tools in support of proteomics research has set the stage for protein and peptide analysis to emerge in the clinical lab. This transition is aided by an emphasis on translational research wherein one aim is the focused conversion of laboratory discoveries into useful clinical tools. Mass spectrometry (MS) is now being driven into service in the clinical laboratory as both a research tool and analytical platform on which accurate and precise diagnostic measurements of proteins and peptides can be made.

Mass spectrometry is not a newcomer to the clinical lab. The implementation of metabolite screening for inborn errors of metabolism is a tremendous success story and established the credibility of mass spectrometry in the diagnostics field in the early 1990's.[1] Since that time, many commercial and hospital labs have adopted MS tools for the quantitative analysis of therapeutic drugs, drugs of abuse, and steroids.[2,3] The benefits of these approaches have been well documented, with improved selectivity, precision and accuracy affording improved clinical utility of diagnostic results. In addition, mass spectrometry-based assays can readily be automated, providing improvements in productivity compared to manual assays that are difficult to implement on robotic platforms.

While the analysis of polypeptides in the clinical lab has traditionally been performed using immunochemical techniques, it appears that mass spectrometry is poised to make significant inroads in this area by bringing a new level of

The author has nothing to disclose.
Quest Diagnostics - Nichols Institute, Research and Development, 33608 Ortega Highway, San Juan Capistrano, CA 92690, USA
E-mail address: cory.e.bystrom@questdiagnostics.com

Clin Lab Med 31 (2011) 397–405
doi:10.1016/j.cll.2011.07.002
0272-2712/11/$ – see front matter © 2011 Elsevier Inc. All rights reserved.

performance to the determination of these analytes. First generation MS-based assays for clinically relevant proteins and peptides will likely to be offered as replacements for immunoassays with known limitations in analytical performance. However, future iterations will likely see the emergence of new markers—discovered by MS-based proteomics approaches, validated using MS-based quantitative methods applied in clinical studies, and finally offered to customers as FDA-approved or laboratory-developed tests (LDT) run on mass spectrometers in reference labs and hospitals around the world.

PROTEOMICS AND THE POST-GENOMIC ERA

The word proteome, a combination of the words *prote*in and gen*ome*, was coined to describe the entire complement of proteins expressed by a genome, cell, tissue, or organism.[4] From this concept, the practice of proteomics emerged as the holistic measurement of these proteins. In these early days, the scientific necessity of such studies was unclear in light of the dramatic progress being made toward the completion of the human genome and the rise in microarray-based transcriptome research. This changed when several key studies highlighted the fact that the intrinsic value of proteomic research demonstrating transcript levels did not correlate well with protein abundance.[5,6] This strongly indicated that proteomics was a necessary and, perhaps, primary element of any biological research effort seeking to understand the fundamental mechanisms of disease. Due to the fact that any disease state could potentially be characterized as a product of aberrant biochemical processes driven by changes in protein structure, abundance, and function, the promise and utility of proteomics as a tool that could be applied to biomarker discovery was apparent.

The clinical research community had long appreciated the potential for proteins and peptides in biological fluids to reveal a sign of disease, essentially practicing very low throughput proteomics years before the phase was ever coined. However, the identification and validation of protein biomarkers was known to be laborious and often delivered markers with modest specificity and sensitivity. Indeed, the limitations of widely accepted markers such as prostate-specific antigen (PSA) and cancer antigen 125 (CA-125) are representative of markers that continue to attract debate regarding appropriate use and clinical utility.[7–9] With this background, the burgeoning promise of the emerging proteomics field was to facilitate "unbiased" and high throughput biomarker discovery. It seemed reasonable that direct comparison of the proteomes of samples from healthy and diseased samples would unambiguously reveal differences in protein abundance mechanistically correlated to disease. In addition, it was proposed that protein biomarkers could ultimately be combined (multiplexed) in order to overcome the challenge of finding a single biomarker with high specificity and sensitivity. These innovations appeared to provide a dramatically improved path toward biomarker discovery that would provide a robust pipeline of new markers useful in the treatment and diagnosis of disease. The promise of these concepts drew the attention of the research community and many labs adopted proteomics technology.

TECHNICAL APPROACHES TO BIOMARKER DISCOVERY

In the early development of proteomics approaches, 2-dimensional polyacrylamide gel separations (PAGE), which coupled both isoelectric focusing- and molecular weight-based separations, were common and provided an entry point to doing comparative and quantitative proteomics.[10–12] With the ability to reproducibly separate hundreds to thousands of proteins and protein isoforms on a single gel,

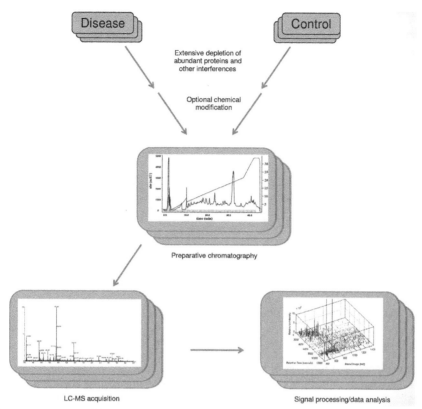

Fig. 1. A prototypical case-control biomarker discovery workflow. (*From* Anderson NL, Anderson NG. The human plasma proteome: history, character, and diagnostic prospects. Mol Cell Proteomics 2002;1:845–67. Erratum in Mol Cell Proteomics 2003;2:50; with permission.)

comparisons of control and experimental gels provided the opportunity to observe changes in protein abundance across a protein constellation. The central role of the mass spectrometer in this analytical pipeline was established through elegant work that demonstrated that data derived from mass analysis of peptides derived from enzymatically digested proteins could be interpreted to rapidly identify proteins without any a priori knowledge.[13] When coupled with microscale protein chemistry techniques, differentially expressed proteins could be identified at an unprecedented rate. Today, gel-centric proteomics approaches have been largely supplanted by highly automated liquid chromatography approaches that have reduced or eliminated the need for manual methods that were highly dependent on operator skill.[14,15]

A typical workflow for proteomic biomarker discovery in its current conception (**Fig. 1**) would start with the collection of patient samples believed to be representative of both disease and healthy states. These samples would be split to generate technical replicates and subsequently processed in parallel. Pre-analytical sample preparation might include steps to reduce interferences or eliminate the most abundant proteins, followed by digestion with trypsin to yield a complex mixture of peptides. This pool of peptides would then be analyzed by mass spectrometry after separation by one or more rounds of chromatography to collect a data set consisting of mass and intensity

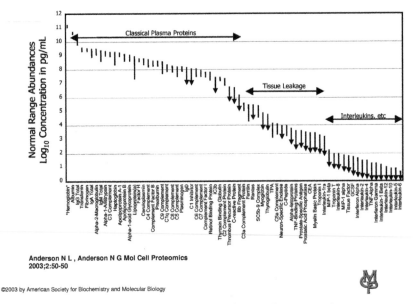

Anderson N L , Anderson N G Mol Cell Proteomics
2003;2:50-50

Fig. 2. The wide dynamic range of some clinically relevant plasma proteins.

measurements as a function of separation time. The raw data is then processed to align 3D contour data (time, intensity, m/z) between samples and then probed for statistically significant relative differences in intensity, with the assumption that these peptide differences may be biologically relevant to the disease state.[16,17] Peptides of interest are then converted to protein identities by a second (or concurrent) round analysis, where the mass spectrometer is used to fragment selected peptides to derive amino acid sequences that are matched to a database of known proteins. Variations of this highly developed platform for differential analysis have been applied extensively for both basic research and clinical discovery proteomics.

As proteomics technologies were being developed, the limitations of the experimental tools became apparent as researchers faced the "protein abundance problem." When considering the dynamic range of protein abundances in samples such as serum, it is no surprise that the overwhelming abundance of a relatively small number of proteins makes it exceedingly difficult to observe low-level proteins. The estimated dynamic range of protein abundances in human serum samples is between 10,000 million and 1 million million, representing the difference between serum albumin, present at approximately 40 mg/mL, down to cytokines and chemokines that circulate at low pg/mL levels (**Fig. 2**). With analytical instrumentation that typically provides 3 to 4 orders of linear dynamic range, the need to "divide and conquer" a proteome by fractionation, depletion, or affinity purification was recognized.[18]

Given that observation of the entire proteome in a single experiment was impossible to achieve, the analysis of subproteomes has become standard procedure. In the field of clinical proteomics, the analysis of low abundance serum proteins and the serum "peptidome" are the 2 major subproteomes that have been examined. The distinction between these 2 proteomes lies in molecular weight. The peptidome fraction is generally considered as the fraction of species less than 15 kDa, and the proteome as those species that are larger. The peptidome has been an attractive target for investigation as a potentially rich pool of bioactive peptides (insulin, Angiotensin II, ACTH), largely free of the overwhelmingly abundant serum proteins.

Despite the tremendous improvement in proteomics technology, the exceedingly high expectations of the degree to which proteomics would deliver biomarkers to the clinic have yet to be realized. Critics have pointed out that support for proteomics has largely been driven by discovery of new putative markers rather than considering the entire research and development process from discovery to commercial in-vitro diagnostic, which includes considerable costs in validation and clinical studies necessary to bring new tests to market.[19,20] In addition, first generation approaches (2D gel approaches) and second generation (low sensitivity Liquid chromatography–mass spectrometry and chip-based platforms) were likely to have insufficient sensitivity to probe the proteome with enough depth to discover robust markers. Finally, the intrinsic complexity of serum, plasma, and tissue is now fully appreciated and reflected in the workflows now employed in most labs.

Proteomics has clearly earned a place in the basic research community. Ground-breaking studies continue to emerge that reveal the complexities of cellular signaling networks, protein-protein interactions, and metabolic pathways. However, the ultimate output of clinical proteomics research has yet to be realized. Research continues to illuminate proteomic differences between tissue and/or serum from healthy and diseased individuals, and candidate markers are being evaluated. Extensive work remains to clinically validate candidate markers and commercialize new tests. Given the disappointments of the past, the analysis of proteins and peptides is emerging in clinical labs, even in the absence of a robust pipeline of new markers.

PROTEINS IN THE CLINICAL LAB

Richard Bright is largely credited for making the association between elevated levels of protein in urine with kidney dysfunction. Given that his work was carried out in 1836, it is impressive to consider that the key role that proteins would play in diagnostics was being established just before they were identified as being chemically unique species. The earliest work on renin was reported nearly 100 years ago using animal bioassays to examine the vasopressive activity of kidney extracts. Fortunately, the crude tools of animal assays and urine coagulation have been supplanted by more modern techniques, none probably as important as the radioimmunoassay technology developed in the 1960's.[21] This immunological method allowed for quantitative measurements of unprecedented sensitivity and selectivity in complex matrices, dramatically enabling the intensive investigation of peptide hormones that circulated at very low levels. This allowed for tremendous growth in the understanding and extensive elaboration of classical endocrine pathways (ie, naturietic regulation, growth regulation). Subsequently, quantitative analysis of many peptide hormones became clinically useful tools and gave rise to widespread use of immunoassay in the diagnostic laboratory.

All immunoassays rely on the unique ability of an antibody to recognize a very specific protein structural feature. This antigen-antibody interaction is coupled to some mechanism that allows for a measurable signal to be generated, frequently an optical response derived from fluorescence or chemiluminescence. This simple assay format has proven to be robust, sensitive, and amenable to automation, providing routine analyses for proteins that circulate at the picogram level. Despite the enormous specificity of antibodies, cross-reactivity and interference are well-known issues that can affect accuracy of a quantitative measurement. While assays are carefully developed to avoid these problems, the enormous heterogeneity of clinical samples taken from humans provides opportunities for unexpected behavior that may have serious consequences.

EMERGING COMPLEXITY OF DIAGNOSTICALLY RELEVANT PROTEINS AND PEPTIDES

Despite the analytical power of immunoassay, ambiguities can arise as a function of protein heterogeneity. Antibodies recognize a 6 to 10 amino acid epitope of a protein as a proxy for the entire molecule. This presents two potential limitations. First, that closely related proteins and peptides can interfere and, second, that epitope loss due to minor degradation or modification will be reflected as decreased abundance of the whole protein even when the majority of the protein remains intact. Together, this means that immunoassays generally do not reflect the complete picture of the target protein in a sample. In many cases, missing information may be inconsequential or have no impact on a quantitative analysis, but evidence is accumulating that information with clinical utility is being lost without a comprehensive analysis of protein heterogeneity.[22]

Mass spectrometry provides an analytical and research tool capable of addressing these issues. The utility of the mass spectrometer to detect and characterize protein structural features while proving quantitative information is derived from the fact that the instrument directly detects molecules and/or molecular fragments and can do so in multiplex. Applying mass spectrometry then allows for the full constellation of proteins to be interrogated in detail. Several groups have used mass spectrometry tools to gain a better understanding of the behavior of clinically relevant analytes that are traditionally monitored by immunoassay.

Lopez and coworkers used a combination of mass spectrometry techniques to interrogate the heterogeneity of parathyroid hormone (PTH), a peptide hormone used diagnostically to diagnose hypo/hyperparathyroidism, hypercalcemia, and renal failure.[23] This study was particularly interesting given the proteolytic processing required for pre-proparathyroid hormone to be processed into what is considered the bioactive form of the peptide hormone (PTH 1-84). Other truncated, and biochemically active forms of PTH have been identified and, while most modern assays examine some of the known forms of PTH, this study revealed a number of new variants and provided an analytical approach for quantitative measurement that would be exceptionally difficult to reproduce using antibodies.

In a recent report, the development of a mass spectrometry-based assay for plasma renin activity revealed a subpopulation of patient samples with very high peptidase activity.[24] Since the plasma renin activity assay relies on the accumulation of Angiotensin I over time, the presence of peptidases was shown to have a dramatic impact on the measured plasma renin activity. Employing the unique ability of the mass spectrometer to multiplex analytes, two internal standards were employed; one to monitor the peptidase activity, and one to control for pre-analytical variability. This allowed both renin activity and peptidase activity to be determined concomitantly so that samples with high peptidase activity could be identified before reporting of an abnormally low result.

A particularly interesting case is the analysis of brain naturietic peptide-32 (BNP-32), used in the clinic as an indicator of congestive heart failure. Testing for this peptide is often carried out in the clinic using point-of-care tests to rapidly diagnose cardiovascular events. Upon analysis of patient samples with high BNP-32 levels as determined by a point-of-care test by mass spectrometry, the authors came to the surprising conclusion that no BNP-32 was actually present.[25] This result highlights the nonspecific nature of some immunoassays, strongly suggesting that the point-of-care test responds to some other form of BNP. While the discrepancy noted for BNP-32 does not necessarily reduce the clinical value of the result generated using the point-of-care test, it does indicate the role that mass spectrometry can play in defining the species that give rise to a immunological signal and, perhaps, indicates that primacy mass spectrometry offers as a tool for quantification of specific chemical species.

In each of these three cases, it is clear that mass spectrometry of proteins and peptides is already having an impact close to the clinical lab and highlights the complexity of diagnostic protein analysis that is, perhaps, currently underappreciated.

LOOKING TO THE FUTURE

Mass spectrometry technology continues to improve at a rapid pace, with continued improvements in sensitivity and robustness. In particular, the use of accurate mass measurements, coupled with the capability to perform direct analysis on intact proteins, is an emerging area that offers new opportunities to examine the relationships between protein post-translational modifications and disease states.[26,27] This technology may also emerge as the mass spectrometry approach that can replace antibody-based immunoassays, without the need for complex pre-analytical sample preparation currently required for many protein assays.

Mass spectrometry will also grow in importance as a preferred platform on which to standardize clinical protein and peptide assays. Discordant measurements of proteins and peptides dependent on antibody and detection method have been a point of concern, especially where these differences confound longitudinal analysis of data where diagnostic measurements may have been made using multiple platforms. In such cases, agreement on reference materials and the use of mass spectrometry are likely to be exceptionally valuable. Methods for important protein analytes like insulin, hemoglobin A1C, c-reactive protein, and c-peptide have been developed, and several have been proposed as reference methods.[28-31]

Over the next several years, it is likely that the current trends will continue, with mass spectrometry methods for protein and peptide analytes demonstrating the practicality of mass spectrometry approaches rather than breaking new ground with novel analytes and new diagnostic tests. In many of these cases, the mass spectrometry methods are likely to offer substantial improvements over conventional methods, or will reveal formerly unappreciated aspects of assay performance. With this experience in hand, clinical chemists will be in a strong position to evaluate and bring next generation diagnostic tools to the clinic.

REFERENCES

1. Chace D, Millington D, Terada N, et al. Rapid diagnosis of phenylketonuria by quantitative analysis for phenylalanine and tyrosine in neonatal blood spots by tandem mass spectrometry. Clin Chem 1993;39:66-71.
2. Dooley KC. Tandem mass spectrometry in the clinical chemistry laboratory. Clin Biochem 2003;36:471-81.
3. Ekman R, Silberring J, Westman-Brinkmalm A, et al, editors. Mass Spectrometry. Hoboken (NJ): John Wiley & Sons, Inc.; 2008.
4. Wilkins M. Proteomics data mining. Expert review of proteomics. 2009;6:599-603.
5. Griffin TJ, Gygi SP, Ideker T, et al. Complementary profiling of gene expression at the transcriptome and proteome levels in saccharomyces cerevisiae. Mol Cell Proteomics 2002;1:323-33.
6. Hack CJ. Integrated transcriptome and proteome data: the challenges ahead. Brief Funct Genomic Proteomic 2004;3:212-9.

7. Vickers AJ, Till C, Tangen CM, et al. An empirical evaluation of guidelines on prostate-specific antigen velocity in prostate cancer detection. J Natl Cancer Inst 2011;103:462–9.
8. Sevinc A, Adli M, Kalender ME, et al. Benign causes of increased serum CA-125 concentration. Lancet Oncol 2007;8:1054–5.
9. Buys SS, Partridge E, Black A, et al. Effect of screening on ovarian cancer mortality: the Prostate, Lung, Colorectal and Ovarian (PLCO) cancer screening randomized controlled trial. JAMA 2011;305:2295–303.
10. Rabilloud T. Two-dimensional gel electrophoresis in proteomics: old, old fashioned, but it still climbs up the mountains. Proteomics. 2002;2:3–10.
11. Rabilloud T, Chevallet M, Luche S, et al. Two-dimensional gel electrophoresis in proteomics: past, present and future. J Proteomics 2010;73:2064–77.
12. Westermeier R, Schickle H. The current state of the art in high-resolution two-dimensional electrophoresis. Arch Physiol Biochem 2009;115:279–85.
13. Henzel WJ, Watanabe C, Stults JT. Protein identification: the origins of peptide mass fingerprinting. J Am Soc Mass Spectrom 2003;14:931–42.
14. Wolters DA, Washburn MP, Yates JR 3rd. An automated multidimensional protein identification technology for shotgun proteomics. Anal Chem 2001;73:5683–90.
15. Link AJ, Eng J, Schieltz DM, et al. Direct analysis of protein complexes using mass spectrometry. Nat Biotechnol 1999;17:676–82.
16. Smith CA, Want EJ, O'Maille G, et al. XCMS: processing mass spectrometry data for metabolite profiling using nonlinear peak alignment, matching, and identification. Anal Chem 2006;78:779–87.
17. Pluskal T, Castillo S, Villar-Briones A, et al. MZmine 2: modular framework for processing, visualizing, and analyzing mass spectrometry-based molecular profile data. BMC Bioinformatics 2010;11:395.
18. Anderson NL. The human plasma proteome: history, character, and diagnostic prospects. Mol Cell Proteomics 2002;1:845–67.
19. Rifai N, Gillette MA, Carr SA. Protein biomarker discovery and validation: the long and uncertain path to clinical utility. Nat Biotechnol 2006;24:971–83.
20. Carr SA, Anderson L. Protein quantitation through targeted mass spectrometry: the way out of biomarker purgatory? Clin Chem 2008;54:1749–52.
21. Yalow R, Berson S. Immunoassay of endogenous plasma insulin in man. J Clin Invest 1960;39:1157–75.
22. Borges CR, Oran PE, Buddi S, et al. Building multidimensional biomarker views of type 2 diabetes based on protein microheterogeneity. Clin Chem 2011;57:719–28.
23. Lopez MF, Rezai T, Sarracino DA, et al. Selected reaction monitoring-mass spectrometric immunoassay responsive to parathyroid hormone and related variants. Clin Chem 2010;56:281–90.
24. Bystrom CE, Salameh W, Reitz R, et al. Plasma renin activity by LC-MS/MS: development of a prototypical clinical assay reveals a subpopulation of human plasma samples with substantial peptidase activity. Clin Chem 2010;56:1561–9.
25. Hawkridge AM, Heublein DM, Bergen HR, et al. Quantitative mass spectral evidence for the absence of circulating brain natriuretic peptide (BNP-32) in severe human heart failure. Proc Natl Acad Sci U S A 2005;102:17442–7.
26. Parks BA, Jiang L, Thomas PM, et al. Top-down proteomics on a chromatographic time scale using linear ion trap fourier transform hybrid mass spectrometers. Anal Chem 2007;79:7984–91.
27. Calligaris D, Villard C, Lafitte D. Advances in top-down proteomics for disease biomarker discovery. J Proteomics 2011;74:920–34.

28. Fierens C, Thienpont LM, Stöckl D, et al. Quantitative analysis of urinary C-peptide by liquid chromatography-tandem mass spectrometry with a stable isotopically labelled internal standard. J Chromatogr A 2000;896:275–8.
29. Kippen AD, Cerini F, Vadas L. Development of an isotope dilution assay for precise determination of insulin, C-peptide, and proinsulin levels in non-diabetic and type II diabetic individuals with comparison to immunoassay. J Biol Chem 1997;272: 12513–22.
30. Kaiser P, Ackerboom R, Ohlendorf R, et al. Liquid chromatography-isotope dilution-mass spectrometry as a new basis for the reference measurement procedure for hemoglobin A1c determination. Clin Chem 2010;56:750–4.
31. Kilpatrick EL, Bunk DM. Reference measurement procedure development for C-reactive protein in human serum. Anal Chem 2009;81:8610–6.

Mass Spectrometry for Clinical Toxicology: Therapeutic Drug Management and Trace Element Analysis

Alan L. Rockwood, PhD, DABCC*, Kamisha L. Johnson-Davis, PhD, DABCC

KEYWORDS
- Mass spectrometry • Ionization • Toxicology
- Clinical chemistry • Therapeutic drug monitoring
- Therapeutic drug management

Mass spectrometry is rapidly expanding its role in clinical chemistry. To date, this growth has primarily been in the area of small molecule analysis, though the field now seems set for a rapid growth of protein and peptide analysis. Small molecule analysis applications can be grouped roughly into the analysis of endogenous compounds and the analysis of exogenous compounds. The analysis of exogenous compounds is largely the domain of toxicology, and this is the focus of this article.

To limit the scope and, therefore, the length of this article, we focus primarily on 2 general topics within clinical toxicology: therapeutic drug monitoring and trace element analysis. While these 2 topics do not cover the full breadth of mass spectrometry applications in clinical toxicology, omitting, for example, clinical drugs-of-abuse testing and applications supporting poison control programs, we feel that they, nevertheless, provide a fairly representative overview of mass spectrometry in the clinical toxicology lab.

TECHNOLOGY AND TECHNIQUE

As an aid to those readers who are less familiar with mass spectrometry technology, we also provide a brief tutorial on the technology of mass spectrometry. This section can also be read as an aid in understanding this article and other articles within this volume.

Department of Pathology, University of Utah School of Medicine and ARUP Laboratories, 500 Chipeta Way, Salt Lake City, UT 84065, USA
* Corresponding author.
E-mail address: rockwoal@aruplab.com

Clin Lab Med 31 (2011) 407–428
doi:10.1016/j.cll.2011.07.003
0272-2712/11/$ – see front matter © 2011 Elsevier Inc. All rights reserved.

labmed.theclinics.com

Fundamental Concepts

The 2 most fundamental concepts in mass spectrometry are that mass spectrometers only deal with ions and that mass spectrometers separate ions according to their mass-to-charge ratio. These concepts are crucial to the understanding of virtually all aspects of both the instrumentation of mass spectrometry and the interpretation of mass spectra. Therefore, the implications of these concepts are far-reaching.

Mass spectrometers depend on electric fields to control the trajectories of molecular-sized species being analyzed within the instrument. Trajectories of charged particles (ions) can be controlled by electric and/or magnetic fields. However, the trajectories of neutral molecules are unaffected by uniform electric and magnetic fields, and are relatively unaffected by nonuniform fields, particularly given the magnitude of the field gradients that can be generated under laboratory conditions. This difference in response to electric and magnetic fields between ions and neutral molecules is the fundamental reason why mass spectrometers can only be used directly for the analysis of ions. A second important, though less fundamental, reason that mass spectrometers are limited to the analysis of ions is that most detectors used in mass spectrometers are either sensitive only to ions, such as faraday cups, or work best when detecting ions, such as electron multipliers.

The force exerted by a uniform electric or magnetic field on a charged particle is proportional to its charge. In the case of a magnetic field, the force is also proportional to the velocity of the particle. Since, by Newton's laws of motion, acceleration is proportional to force and inversely proportional to mass, for a given combination of electric and magnetic fields, the trajectory of an ion in a mass spectrometer depends on the mass-to-charge ratio of the ion, also known as m/z.

Because the trajectory of an ion in the mass spectrometer depends not on mass but on m/z, the term "mass spectrometer" is a misnomer. If one were to use a naming convention in strict accordance with the function of the instrument, we would call these devices "mass-to-charge ratio spectrometers." We do not intend to correct this little glitch in terminology, but it is important to correctly understand the function of a "mass spectrometer" in order to avoid certain conceptual errors that might otherwise occur.

The implications of the fact that mass spectrometers only deal with ions and only measure m/z are by no means trivial. They are fundamental to the interpretation of a mass spectrum. For example, barium has 7 isotopes. Doubly charged barium (Ba^{++}) would appear in a mass spectrum as 7 isotope peaks at the following mass-to-charge ratios: m/z 65.0, 66.0, 67.0, 67.5, 68.0, 68.5, and 69.0. If one were to incorrectly assume that m/z represented mass, one would underestimate the atomic mass of barium by a factor of 2. Furthermore, the Ba^{++} peaks at m/z 66, 67, and 68 have the same nominal masses as 3 of the isotope peaks of zinc, appearing as a singly charged ion (Zn^{+}). Thus, doubly charged barium could be a potential interfering species in the analysis of Zn+ by mass spectrometry.

Putting these concepts together, we conclude that the block diagram of a mass spectrometer must contain at least 3 main components: an ion source, a m/z analyzer (usually referred to as a "mass analyzer"), and a detector. To these 3 main components one can add a sample introduction device and components to process the signal generated by the detector, such as amplifiers, digitizers, and computers.

Ion Source

Let us now briefly discuss the technology of the principal components of a mass spectrometer, starting with the ion source. Mass spectrometers only detect ions. In most cases, an analyst is interested in analyzing for a compound that is nominally non-ionized. This natural mismatch between the sample and the instrument must be bridged in some way. To bridge this gap, mass spectrometers contain an ion source whose function is to convert the non-ionized molecules in the sample into ions for analysis by the mass spectrometer. Many ion source technologies have been developed during the nearly 100-year history of mass spectrometry. However, 4 ion sources dominate clinical applications of mass spectrometry. They are electron ionization (EI), electrospray ionization (ESI), atmospheric pressure chemical ionization (APCI), and inductively coupled plasma ionization (ICP).

Two of these ion sources (EI and ICP) are "hard," meaning that molecules are both ionized and broken into smaller fragments. EI is the better known and more widely used of the two. An EI ion source consists of a small evacuated chamber into which the sample is introduced in gaseous or vapor form. The pressure in the ion source is kept low, typically so that there are few if any collisions between analyte molecules and background gas. An "electron gun" accelerates a stream of electrons and directs it into the vaporized sample. Some of these electrons collide with sample molecules, causing the ejection of an electron and leaving behind a positively charged ion, typically a polyatomic ion. One can think of this as a "1 electron in, 2 electrons out" process.

In the first instant after ionization, the chemical structure of the molecule is intact except for the loss of an electron. However, ionization in an EI source is an energetic process and the resulting ion is usually unstable, in most cases breaking into smaller polyatomic fragments, one of which is charged and, therefore, amenable to mass spectral analysis. This fragmentation process is statistical but not completely random, meaning that the initially formed ion tends to break into a reproducible pattern. Thus, in most cases, an EI mass spectrum is composed primarily of fragment ions, and the abundance pattern of the fragment ions is compound-dependent and relatively reproducible. Therefore, mass spectra from an EI source can be used as a kind of "fingerprint" for identifying a compound matching the unknown mass spectrum to previously recorded mass spectra in a library of mass spectra.

For many decades, EI was the ion source most widely used for mass spectrometry. EI is compatible with the gas flow rates from capillary gas chromatography, providing a good way to couple a gas chromatograph to a mass spectrometer. However, an EI ion source requires that the sample be volatile, or at least semi-volatile. Heating of the sample is often required and, as a practical matter, not all samples can be volatilized, either because they are thermally labile or because they are essentially nonvolatile even at elevated temperatures. Often these "difficult" compounds are highly polar and/or present as preformed of ions, and/or of high molecular weight, all of which tend to cause low volatility. Many—possibly even most—clinically interesting compounds fall into this category. Hence, the role of EI in clinical mass spectrometry is limited to a relatively narrow range of applications.

Inductively coupled plasma ionization can be thought of as the ultimate "hard" ion source because the mass spectra are composed largely of atomic ions rather than polyatomic ions. An ICP ion source is composed of several components. The first is a nebulizer that generates a fine spray from a liquid sample. After nebulization, the sample stream becomes a plasma (a hot, highly ionized gas) generated in an argon gas stream by coupling a high-powered radio frequency (RF) power source into the plasma using an electrically driven coil. The coil is an inductor, hence the name "inductively coupled plasma" for this ion source.

The plasma in an ICP torch is at atmospheric pressure, and its temperature is comparable to that of the surface of the sun. At these temperatures, almost all

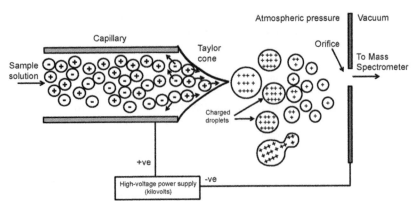

Fig. 1. Components of an electrospray ion source.

molecules that enter the plasma become dissociated to atoms, and a large fraction of the atoms become ionized. A portion of the plasma is drawn into the vacuum system of the mass spectrometer through a small (sub-mm) hole. The mass spectrometer then separates the sample by m/z. Because molecules that enter the plasma become completely dissociated, chemical structure information is lost in the process, and ICP is, therefore, best used as an ion source for elemental analysis. In the clinical laboratory, ICP is used primarily as an ions source for trace element analysis.

At the time of this article's writing (2011), ESI is the most widely used ion source in mass spectrometry and, for several reasons, this popularity carries over into clinical mass spectrometry. Unlike EI, which requires that the sample be volatile or, at least, semi-volatile, ESI is useful for analysis of a wide variety of nonvolatile compounds. This opens the door for the application of mass spectrometry to a greater range of clinical applications.

Electrospray ionization requires that the compound(s) of interest be soluble in a solvent compatible with the ESI process. The best solvents tend to be small, polar, organic compounds, such as methanol or acetonitrile, though other solvents are sometimes used. Water can be used as a solvent, though ESI generally works better for solvents such as methanol or acetonitrile, and water is, therefore, more often used as one component of a mixed solvent rather than as the sole solvent.

An electrospray ion source consists of several components at atmospheric pressure, followed by a sampling orifice for transporting ions into the vacuum system of the mass spectrometer (**Fig. 1**).

The sample, in the form of a solution of the compounds of interest, elutes from the end of a capillary. A voltage measured in kilovolts is applied to the solution. Electrical forces on the solution causes a spray of highly charged droplets to be emitted from the solution. In some cases, the nebulization process is aided by a flow of gas, somewhat reminiscent of the operation of a carburetor in an automobile.

The best hypothesis of what follows is that the highly charged droplets begin to evaporate and, as the evaporating droplets become smaller, the electrical forces that tend to destabilize the droplet eventually overcome the surface tension forces that stabilize the droplet and, at the "Rayleigh limit," the droplet becomes unstable and breaks into smaller droplets. This process is often aided by the introduction of a heated gas stream. Eventually, analyte ions leave the solution, probably aided by electrostatic forces, and are entrained in a flow of gas entering through an orifice into the vacuum system of the mass spectrometer.

Unlike an EI source, in which ions are generated from vaporized sample, an ESI ion source conceptually relies on the existence of preformed ions (ie, the compounds of interest may already exist in the form of ions in solution). The ESI source is simply a device to transfer these preformed ions into the gas phase and then into the mass spectrometer. Therefore, ESI depends significantly on the solution phase chemistry.

In most cases, the solution phase ions to be observed in ESI mass spectrometry are a product of acid-base chemistry. For example, basic compounds tend to attach a proton to make positive ions in solution. These are most favorably observed using a low pH solvent while running positive ion mode ESI, and the expected ion to be observed in the mass spectrum is the analyte molecule with the addition of one or more protons. Conversely, acidic compounds are most favorably observed using a high pH solvent while running in negative mode ESI, and the expected ion to be observed in the mass spectrum is the analyte molecule modified by the loss of one or more protons. There are exceptions to these general trends. For example, metal or ammonium adduct ions may sometimes be observed, and ions formed by redox processes have also been observe.[1]

Electrospray ionization is particularly notable for 2 properties. First, it is a very soft ionization method. Fragment ions are generally absent and, when present, they are usually the result of processes independent of the ionization process itself, such as collisionally activated dissociation of ions in the differential pumping region of the electrospray/mass spectrometer interface. Second, ESI exhibits a remarkable propensity for producing multiply charged ions, particularly for high molecular weight compounds such as peptides, proteins, and other biopolymers. For example, ubiquitin has a molecular mass of 8564.47 Da. In a positive ion, ESI is observed not at an m/z of 8565.47, which is what one would expect if a single proton were attached to the ubiquitin molecule to form a singly charged ion, but rather as a series of peaks, such as m/z 659.81, 714.71, 779.60, 857.45, 952.62, and 1071.57. Each peak corresponds to a different number of proton attachments, resulting in a different charge and a different m/z (**Fig. 2**). For example, the peak at m/z 659.81 corresponds to the attachment of 13 protons, and the charge is +13. Although the peaks for successive charge states have nearly the same masses, being 8577.56, 8576.56, 8575.55, 8574.54, and 8573.54 Da respectively, with successive peaks differing in mass by only the mass of a proton, they differ substantially in m/z primarily because of the difference in charge, being +13, +12, +11, +10, +9, and +8 respectively.

Perhaps the most attractive feature of the ESI ion source is that it operates directly on a liquid phase sample, so it provides a natural interface between a liquid chromatograph and a mass spectrometer (LC-MS). Because gas chromatography-mass spectrometry (GC-MS) requires that the sample be volatile and stable (usually at elevated temperatures), and because many compounds do not fulfill those requirements, LC-MS is useful for a wider range of samples than GC-MS.

Atmospheric-pressure chemical ionization (APCI) is the final ion source to be discussed here. The components include a pneumatically assisted nebulizer, a heated gas stream, a corona discharge needle, and a sampling aperture for transporting ionized samples into the vacuum system of a mass spectrometer. An APCI ion source looks a lot like some ESI ion sources, and some instruments are designed with dual capability in the same source. The main mechanical/electrical difference between an ESI source and an APCI source is that high voltage is not applied to the solution in APCI but, rather, is applied to the corona discharge needle.

At a mechanistic level, ionization in APCI does not take place directly from the solution phase as in ESI but, instead, analyte molecules are formed by vaporization as the droplets in the spray evaporate, followed by ionization through a series of ion

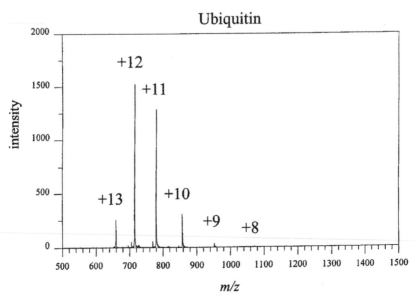

Fig. 2. Ubiquitin's molecular mass.

molecule reactions. The ion molecule reactions are initiated by the corona discharge and may involve a several step mechanism in which vaporized solvent molecules play a role. Atmospheric-pressure chemical ionization and ESI have a broad range of overlapping applications. However, compared to ESI, APCI tends to be more useful for molecules that are smaller and/or less polar than those best adapted to ionization by ESI.

Other ion sources are available as well, such as chemical ionization (CI), matrix-assisted laser desorption ionization (MALDI), and others. However, because these have, so far, had little impact in clinical mass spectrometry, they will not be discussed here.

Mass Analyzers

There is a wide array of *m/z* analyzer technology available, which we will loosely refer to as "mass analyzers". However, quadrupole mass analyzers dominate. A quadru-pole mass spectrometer consists of 4 accurately machined conductive rods arranged in a square array with the rods parallel to each other. The region between the rods can be thought of as an ion channel through with ions pass during mass analysis. Ions from the ion source enter at one end of the array and travel toward a detector located at the other end of the array. Some ions are ejected laterally from the array. Other ions travel the length of the array and are detected by a detector placed at the far end of the array.

For a given set of instrument parameters there is a *m/z* setpoint and passband (Δ *m/z*) that determines which ions are transmitted to the detector and which are not. These are determined by voltages applied to the quadrupole rods. The combined voltage is the superposition of a radio frequency component (RF) and a constant component (DC).

A full description of the physical principals operating in a quadrupole mass spectrometer requires some rather complex mathematics (ie, solutions to the Mathieu equation). However, a qualitative description is as follows: The RF part of the potential causes ions below a certain *m/z* cutoff to have unstable trajectories, and they are thrown out of the rod assembly in a radial direction. Above this cutoff, the RF field tends to prevent loss of ions in the radial direction. However, above a certain *m/z*, the DC part of the potential can overcome the confining effect of the RF field, and this causes ions of high *m/z* to be ejected. The result is that only ions within a certain *m/z* window can travel the length of the quadrupole rod assembly. The center point and width of this window determines the *m/z* setting and resolution of the instrument at any given moment. Resolution is defined as *m/z* of a given peak divided by the peak width in *m/z* units. By changing field settings, the instrument can detect ions of a selected *m/z* with a selectable resolution.

The instrument may be programmed to multiplex through a table of selected *m/z* values (selected ion monitoring, or SIM mode), or the instrument may be programmed to scan smoothly over a range of *m/z* values (scanning mode). Regardless of the mode of operation, at any given instant, the instrument will be monitoring only ions within a Δ *m/z* passband of some selected *m/z* setting.

In general, one can characterize most quadrupole mass spectrometers as having moderate sensitivity in scan mode, excellent sensitivity in SIM mode, moderate resolution (roughly, 1000 < resolution <10,000) and moderate-to-good scan speed. Mass accuracy is moderate, typically of the order of 100 ppm. Mass range is also moderate, typically of the order of 2000 *m/z* units. These instruments feature ease of use and relatively economical pricing.

Another mass analyzer, time-of-flight (TOF), is beginning to have an impact in clinical mass spectrometry. In the simplest conceptual model of this device, at time $t = 0$ ions are all bunched in space with a negligible spread in kinetic energy. The ion packet is then accelerated down the axis of a flight tube and strikes a detector. One can visualize the device as an evacuated pipe, with the ion source at one end and the detector at the other. The flight time from source to detector depends on *m/z* and the electrical potential through which the ions are accelerated, ϕ

$$t = C\frac{\left(\dfrac{m}{z}\right)^{\frac{1}{2}}}{(\phi)^{\frac{1}{2}}}$$

where C depends on geometric factors such as flight tube length.

Neither of the simplifying assumptions regarding spatial distribution of ions and/or kinetic energy distribution of ions is strictly true in practice, and the breakdown of these assumptions can lead to a reduction in resolution. Most modern TOF mass spectrometers incorporate one or more design features to mitigate this loss of resolution, notably orthogonal introduction of ions into the time-of-flight mass spectrometry (TOF-MS) and/or the incorporation of ion mirrors. Consequently, many currently offered TOF mass spectrometers boast resolution specifications in the 10,000 to 100,000 range.

Time-of-flight-mass spectrometry has a number of very attractive features, including very high acquisition speed (some instruments achieving >100 full mass spectra per second stored to disk), high *m/z* range (often >10,000 *m/z* units), excellent mass accuracy (typically <5 ppm), moderate sensitivity for single *m/z* detection and high sensitivity on a full-spectrum basis when compared to a quadrupole MS.

There are a number of other mass analyzer technologies available, including linear ion trap, quadrupole ion trap, ion cyclotron resonance, orbitrap, and magnetic sector. Each has its unique set of advantages and disadvantages. It remains to be seen whether any of these will be widely adopted by clinical laboratories. However, among these, the best candidates for adoption by clinical laboratories are probably the linear ion trap combined with a quadrupole mass spectrometer to form a hybrid tandem MS, and the orbitrap combined with a linear ion trap to form a hybrid tandem MS.

Hyphenated Techniques

In the clinical lab, mass spectrometers are seldom used alone. They are usually used as part of a "hyphenated" analysis. This is based on the concept that analytical specificity can be improved if one can measure more than one physical property of a compound. For example, most mass spectrometers in clinical mass spectrometry are combined with a chromatograph, such as liquid chromatography-mass spectrometry (LC-MS). The chromatograph separates the sample mixture according to retention time. The mass spectrometer then detects each compound according to m/z as it elutes from the chromatograph. Thus, each compound is characterized by 2 physical properties, retention time and m/z. Because retention time is typically not highly correlated with m/z, the chromatographic part of the analysis is "orthogonal" to the mass spectral part of the analysis, and the result is a highly specific analytical scheme.

The combination of gas chromatography with mass spectrometry (GC-MS) has long been used in clinical laboratories but, in recent years, liquid chromatography (LC) has largely supplanted GC as the preferred chromatographic technique for chromatography-mass spectrometry in the clinical laboratory. Reverse phase LC is by far the most popular technique for use with mass spectrometry, and there is an increasing trend toward higher pressure LC technologies. The benefits of higher pressure include shorter run time and/or higher chromatographic resolution.

Another hyphenated technique is to combine 2 mass spectrometers. The first mass spectrometer selects a "precursor" ion (referred to as a "parent" ion in the older literature) according to its m/z. Precursor ions are then broken into smaller pieces, usually by collisions with background gas in a low pressure "collision cell". These pieces are known as "product ions" (referred to as "daughter ions" in the older literature.) A second mass spectrometer then selects product ions according to m/z. This technique is known as MS/MS or tandem mass spectrometry. Despite the nomenclature anomaly of using a "/" rather than a "-", MS/MS is considered a hyphenated technique because it characterizes a compound by 2 physical properties, precursor ion mass and product ion mass.

Most tandem mass spectrometers use triple quadrupole technology. This instrument is composed of 3 quadrupoles, the first and last (Q1 and Q3) being mass-resolving and the middle one (Q2) being tuned to transmit a broad m/z range through the device (ie, it acts as an "ion pipe", ideally transmitting all ions regardless of m/z), though, in reality, there will be a limited, albeit extremely broad, passband. An inert gas at low pressure is introduced into Q2 to serve as a collisions gas. Parent ions collide with background gas molecules. This energizes precursor ions, causing them to dissociate into product ions.

Tandem mass spectrometers may also be of mixed or "hybrid" technology. For example, if the first mass spectrometer is a quadrupole mass spectrometer, and the second mass spectrometer is a time-of-flight mass spectrometer, then the instrument is commonly referred to as a Q-TOF MS. Hybrid instruments are typically designed to make use of the complementary characteristics of different analyzer technologies. For example, a Q-TOF combines the efficient m/z-selective transfer of ions from ion source to collision cell of a quadrupole MS, and the TOF MS provides high resolution, accurate mass, and efficient simultaneous detection of the full mass spectrum of product ions.

Even more popular than LC-MS or MS/MS in the clinical mass spectrometry lab is liquid chromatography-tandem mass spectrometry (LC-MS/MS). This technique provides excellent analytical specificity because it characterizes a compound by 3 physical properties—retention time, precursor ion mass, and product ion mass. For many clinical applications, LC-MS/MS provides a favorable trade-off between the complexity of the technique and analytical specificity; hence, this technique dominates the clinical mass spectrometry lab.

The MS/MS transition is a key concept in tandem mass spectrometry. It is the combination of precursor ion m/z and product ion m/z. For example, in our laboratory we have a method for cortisol that produces a precursor ion of m/z 363 and a product ions of m/z 121, so the MS/MS transition is 363→121. As second product ion from the same parent ion is m/z 97, in which case the MS/MS transition is 363→97. For quantitative analysis of target compounds, one usually sets up a table of MS/MS transitions and performs a multiplexed measurement, cycling repeatedly through the MS/MS transitions in the table. If more than one compound is being analyzed, then the table will include a parent ion m/z for each compound of interest and, for each parent ion selected, there may be one or several product ions in the table. This form of analysis is sometimes known as selected reaction monitoring (SRM), or multiple reaction monitoring (MRM). Other scan functions such as product ion scan, precursor ion scan, and constant neutral loss scan are also possible, but they are less widely used for targeted quantitative analysis in the clinical lab and will not be discussed here.

Technique Notes

Quantitative analysis by mass spectrometry of target compounds is usually done with the aid of an internal standard that is spiked into the sample. The analysis is then based on the ratio of instrument response of the target compound to that of the internal standard. The internal standard should be chemically similar to the target compound so its recovery can mimic that of the target compound during sample preparation and analysis. The ideal internal standard is an isotopically labeled version of the target compound. Unlike most detectors, a mass spectrometer can distinguish between isotopically labeled and normal versions of the target compound and, given that an isotopically labeled compound has nearly identical physical and chemical properties to that of the unlabeled compound (including LC retention time, solubility, derivatization chemistry, ionization efficiency, and so forth), isotopically labeled internal standards will faithfully mimic the target compound during sample preparation and analysis and are, therefore, used whenever possible.

Not all isotopically labeled internal standards are of equal value. For example, deuterated compounds may have subtly different properties than unlabeled compounds, such as retention time shifts or, in some cases, labile labeling (H/D exchange), that may make them less desirable than [13]C- or ^{15}N-labeled compounds. However, cost and availability considerations will often tip the decision toward deuterated internal standards. In the case of inductively coupled plasma mass spectrometry, isotope-based internal standards are less widely used and it is more common to use a low-abundance element for the internal standard, such as yttrium, scandium, gallium, or other suitable choice.

Mass spectrometry, particularly when incorporated as part of a hyphenated technique, is widely praised for its analytical specificity. This is one of the major motivations for converting an assay from immunoassay or enzymatic assay to mass spectrometry. In comparing analytical specificity of mass spectrometry to a typical immunoassay, it is well to consider and contrast the potential interference profiles of the 2 techniques. In the case of immunoassays, the most frequent interfering compounds are structural analogues of the target compound. Molecular weight similarities are of no particular concern, although some structural analogues may, by accident, have the same molecular weight as the target compound. Immunoassays for small molecules may be especially prone to interferences. For example, in our laboratory, drugs-of-abuse screening by immunoassay methods often produce a very high false positive rate that must be resolved by using a different method to confirm the result.

By contrast, interferences in mass spectrometry-based methods are relatively uncommon, particularly if mass spectrometry is embedded in a hyphenated methodology. Rather than structural analogs, interferences in mass spectrometry are based on molecular weight, either of the unfragmented ion or of fragment ions. In the case of tandem mass spectrometry, a compound would need to have the same mass-to-charge ratio as the target compound for both the precursor and the product ion in order to interfere. Compared to many analytical methods, mass spectrometry is a relatively high resolution technique, so, generally speaking, interference is a relatively low probability event. However, even structurally unrelated compounds may sometimes cause unexpected interferences. For example, in our lab we have noted that fenofibrate has potential to interfere with cortisol in MS/MS because both compounds produce a precursor ion at m/z 363, and they both produce an abundant product ion at m/z 121.[2] However, a secondary MS/MS transition for cortisol, m/z 373 → 97, has no corresponding transition in fenofibrate.

This example illustrates 2 important concepts. First, in developing an assay for a target compound, it is generally impossible to screen for all possible interferences, so there is always a possibility that an unexpected interference can occur in an individual patient's sample. Second, mass spectrometry provides an opportunity to check for interferences in each patient sample and flag results that show evidence of interference. The general scheme as follows: One monitors the intensity of at least 2 peaks in the mass spectrum or the MS/MS spectrum. The abundance ratio of these 2 peaks must be within some tolerance band compared to the ratio of an authentic pure standard included in the same batch. If the ratio is too high or too low, then the sample is flagged as containing an interfering compound. Using this scheme, one can, in most cases, avoid reporting results that may be invalid because of interferences.[3]

Terminology Note

When a clinical chemist refers to "specificity", he or she may be referring to 1 of 2 concepts: either the ability to distinguish one compound from another (analytical specificity) or the ability to correctly classify a nondiseased patient as disease-free (clinical specificity). Consequently, one must be careful in the use of terminology to be sure there is no confusion. Context alone is not always sufficient to clarify which type of specificity is under discussion, so it is good practice to use the terms "analytical specificity" or "clinical specificity" to eliminate confusion.

Similarly, one must be careful to avoid confusion between analytical sensitivity (commonly taken to mean a favorable limit of quantitation) from clinical sensitivity (the ability to correctly classify a diseased patient as diseased or nondiseased.)

The term "calibration" is also used in at least 2 ways. Most frequently, when a mass spectrometrist "calibrates" a mass spectrometer, one is referring to calibration of the m/z scale. However, calibration may also refer to preparation of a calibration curve for quantitative analysis (ie, determining the relationship between concentration and instrument response).

TRACE ELEMENT ANALYSIS

Elemental analysis of human samples is an important, though specialized, part of medical toxicology. For example, lead exposure is a well-recognized health hazard. The state of New York considers that, for adults with blood concentrations between 10 and 25 μg/dL, lead is building up in the body and some exposure is occurring. Between 25 and 40 μg/dL, regular exposure is occurring, and there is some evidence of potential physiologic problems.[4] The same reference states that lead poisoning in children is especially dangerous. The US Centers for Disease Control has set a series of cutoff levels for whole blood lead for children, starting at 10 μg/dL, with a graded series of case management recommendations depending on the confirmed concentration of lead in venous blood.[5]

Classical elemental analysis methods based on wet laboratory procedures have long ago been replaced by instrumental methods of chemical analysis. One classical approach, the flame test, foreshadowed the development of modern instrumental analysis techniques. In the flame test, one would introduce the sample into a flame. The presence of light emission of particular colors was associated with the presence of certain elements in the sample, usually metals. This was a form of qualitative analysis. With the addition of certain components, such as monochromators and sensitive light detectors, the flame test evolved into flame emission spectroscopy, which is capable of quantitative as well as qualitative analysis, and is also capable of multi-analyte detection. A further development led to the replacement of the flame with an inductively coupled plasma, yielding ICP-atomic emission spectrometry (ICP-AES).

Taken in a slightly different conceptual direction, it is possible to add a light source to a flame emission spectrometer to measure the intensity of light of a specific wavelength as it passes through the flame. The selected wavelength is matched to the absorption spectrum of atoms of an element chosen for analysis. Absorption of light leads to attenuation of the beam, and this is indicative of the presence of the selected element. The degree of absorption is proportional to the concentration of the element in a liquid sample nebulized into the flame. This technique, flame atomic absorption spectroscopy, was further elaborated by replacing the flame with a graphite furnace, yielding graphite furnace atomic absorption spectroscopy.

Taken in still a different conceptual direction, if one were to delete the light detector from ICP-AES and substitute a mass spectrometer for sampling the plasma, one would have the basis for an inductively coupled plasma mass spectrometer (ICP-MS). The inductively coupled plasma mass spectrometer instrument can generally be characterized as having excellent sensitivity, good selectivity, excellent elemental coverage, high speed for a single analyte, and capability for multi-element multiplexed analysis. A table comparing the relative merits of the technologies for elemental analysis is found in the work by Rockwood and Bakowska.[6] Taken as a whole, ICP-MS has a very favorable balance of advantages/disadvantages relative to other technologies for elemental analysis, its chief drawback being high initial cost of instruments.

The range of elements that can be analyzed by ICP-MS is very broad. For example, our laboratory offers 18 elements by ICP-MS in as many as 5 different matrices (blood, urine, serum, plasma, and tissue), plus 3 elements used exclusively as internal standards. Another major reference laboratory offers 31 elements by ICP-MS in as many as 6 different matrices (blood, urine, serum, plasma, tissue, and red blood cells), with additional matrices (eg, hair, nails, and other tissues or biological fluids) available on a nonroutine basis. The range of elements offered by clinical or toxicology labs is limited more by clinical utility than by the technique itself. For example, one company (EAI) specializing in elemental analysis for clients outside of the clinical testing arena offers analysis for 71 different elements by ICP-MS in a wide variety of matrices.

Sample preparation for ICP-MS in the clinical lab varies with the type of sample. However, for most noncellular fluids, sample preparation follows a dilute and shoot protocol (ie, the addition of an internal standard, usually an uncommon element that is not being targeted in the assay, possibly followed by a dilution step). Samples containing cellular components (whole blood, red blood cells, or tissues) generally require additional processing, such as a nitric acid digestion step.

One of conceptual advantages of ICP-MS is a relative freedom from interferences. However, one must not assume that the technique is interference-free, and considerable attention is paid to dealing with this problem. Interferences can be broken down into 3 main classes: isobaric molecular ions, isobaric elemental interferences from singly charged ions, and doubly charge elemental ions.[7]

It may seem surprising that isobaric molecular ions would be a problem, given the simplified conceptual model of ICP-MS, which has a very hot plasma that breaks molecular species down to the atomic level. Nevertheless, interferences from molecular ions can sometimes be significant. The molecular interferences are generally recombinations of plasma components, such as Ar, H, O, Cl, or other elements. For example, $^{38}ArH^+$ and $^{23}Na^{16}O^+$ can both interfere with $^{39}K^+$, $^{40}Ar^{35}Cl^+$ can interfere with $^{75}As^+$, and $^{16}O^{40}Ar^+$ is a well-known interference against $^{56}Fe^+$. A list of polyatomic interferences has been published by May and Wiedmeyer.[8] Less commonly, elemental isobaric interferences may occur. In this case, an isotope of one element might interfere with an isotope of another element if both isotopes have the same nominal isotopic mass. The most difficult cases would be if there were elemental isobaric interferences against mono-isotopic elements, such as ^{27}Al, ^{55}Mn, or ^{75}As, because, in such cases, there would be no alternative isotope of an element to monitor. Fortunately, there are no such cases. Conversely, there are no cases in which a mono-isotopic element interferes against an isotope of a poly-isotopic element. However, interferences between some poly-isotopic elements are possible. For example, in the unlikely event that one was to select the $^{54}Cr^+$ isotopic peak for the analysis of chromium, $^{54}Fe^+$ could be a potential interference. Similarly, $^{40}Ar^+$, $^{40}K^+$, and $^{40}Ca^+$ represent a set of isotopic peaks that could interfere with each other.

Among all the elemental isotopic combinations possible, there are many cases where elemental isotopic interferences could be possible in principle. Most, however, are unlikely to affect clinical samples analyzed by ICP-MS. For example, the most abundant isotope of Mo^+, $^{98}Mo^+$, has a potential interference from $^{98}Ru^+$. However, Ru is not likely to be found in significant concentrations in a clinical sample, and even if it were, the $^{98}Ru^+$ isotope is a relatively minor (<2%) fraction of total Ru. Furthermore, there are several other isotopes of Mo^+ that could be selected for the analysis instead of $^{98}Mo^+$. A more problematic case would be the analysis of elements producing m/z 40 ions by ICP-MS. In this case, the set of isotopes $^{40}Ar^+$, $^{40}K^+$, and $^{40}Ca^+$ could interfere with each other in various combinations. However, $^{40}K^+$ is a relatively nonabundant isotope, so it is not likely to interfere, leaving $^{40}Ar^+$ as a potential interfering species against $^{40}Ca^+$. A method of dealing with interference by $^{40}Ar^+$ will be discussed later.

Interferences by doubly charged ions is also possible in ICP-MS. The potential interference of Zn^+ by Ba^{++} was given earlier. The relatively low first and second ionization potentials of Ba (in total, less than the first ionization potential of Ar) would favor this. However, it would be unusual for Ba concentrations to be high enough to cause significant interference with Zn. There are, in fact, many potential m/z overlaps between doubly charged atomic ions and singly charged atomic ions (ie, most singly charged ions having nominal m/z between 1 and 119, inclusive). At the low end, only m/z 4, 5, and 8 are free of potential interferences from doubly charged atomic ions and, at the high end, only m/z 106-110, 112, 114, 115, and 118 are free of potential interferences from doubly charged atomic ions. However, as a practical matter, few doubly charged atomic ions are likely to affect clinical analysis, either because the potentially interfering elements are rare in clinical specimens, the particular isotopes for the doubly charged element are uncommon, an element may not form doubly charged ions efficiently, or the isotope of the singly charged ion may be of low abundance and, therefore, of little interest for quantitative analysis.

There are at least 3 methods to deal with interferences in ICP-MS. The first is to use a high resolution mass spectrometer. Marchante-Gayon and colleagues have summarized the resolving power required to resolve interferences in biological samples.[9] A shorter list more focused on interferences of clinical and toxicological importance has been given by Patriarca and colleagues.[10] In most cases, the required resolving power is greater than 10, 000. For example, the mass of $^{56}Fe^+$ and $^{16}O^{40}Ar^+$ differ by 1 part in 2333. Therefore, an instrument offering a baseline resolution of greater than ~2500 would be capable of selecting $^{56}Fe^+$ over $^{16}O^{40}Ar^+$ or, if one were to specify this parameter very conservatively, a resolution of greater than ~5000 would be capable of detecting $^{56}Fe^+$, even in the tail of an $^{16}O^{40}Ar^+$ of much higher abundance. In a few cases, the required resolution is much higher, such as the $^{40}Ca^+/^{40}Ar^+$ pair, which would need a resolution of 193,000. At least one vendor, Thermo Fisher, offers high resolution ICP-MS instruments. These instruments have, so far, found little penetration into the clinical ICP-MS market.

A more popular method of dealing with interferences is to place a low pressure collision cell or reaction cell between the ion source and the mass analyzer. In the case of a reaction cell, a low pressure of a reactive gas, such as hydrogen or ammonia, is bled into the cell. The reactive gas removes interferences via chemical reaction. For example, interference by $^{16}O^{40}Ar^+$ can be removed via the following charge transfer reaction:

$$^{40}Ar^{16}O^+ + NH_3 \rightarrow {}^{40}Ar^{16}O + NH_3^+$$

$^{40}Ar^{16}O$ is electrically neutral and is, therefore, not detected in the mass spectrometer.

In some cases, algebraic formulas can be used to correct a result for the presence of interferences. For example, selenium has several isotopes, the most abundant being ^{78}Se and ^{80}Se. None of the Se isotopes are free from potential interferences by polyatomic species. For example, $^{40}Ar^{40}Ar^+$ can interfere with $^{80}Se^+$ and $^{40}Ar^{38}Ar^+$ can interfere with $^{78}Se^+$. By using the known relative isotopic abundances of the isotopes of Se and Ar, measuring the relative abundance of the m/z 78 and 80 peaks, and solving a set of linear equations, one can correct the results for the interferences. This example is somewhat contrived and simplified because there are many additional potential interferences of Se by other atomic and polyatomic ions. As more interferences are considered, the algebraic equations increase in complexity and the error propagation of statistical variations and other errors through the calculation becomes more of a problem (ie, the problem runs the risk of being mathematically ill-conditioned or even underdetermined). An ill-conditioned system is numerically unstable and therefore difficult to solve. An underdetermined system is not amenable to a unique solution unless additional constraints are applied, such as assuming that some terms are small enough to ignore or performing additional measurements on *m/z* outside of the isotopic cluster of the element of interest. Somewhat similar in spirit, but employing mathematical methods to incorporate additional information into the calculation, Sharp and colleagues[11] have described a Bayesian analysis of inductively coupled plasma mass spectra in the range 46 to 88 Da derived from biological materials. This method can also take into account anomalies in the data due to the mass-dependent response of the mass spectrometer.

Inductively coupled plasma mass spectrometry is generally regarded as a method for elemental analysis because, for the most part, the ICP torch breaks the sample down to the atomic level. Thus, chemical information about the sample is lost. This can be an advantage if one is only interested in the total concentration of an element, regardless of the chemical form of the element. However, in some cases, management of a patient requires knowledge of an element's speciation (ie, the chemical form of the element). For example, the clinical implications of an elevated urine arsenic concentration are drastically different for a patient who has ingested highly toxic inorganic arsenic compounds compared to one who has ingested a seafood meal containing much less toxic organic arsenic compounds. Unfortunately, conventional ICP-MS is unable to distinguish between the various chemical forms of arsenic.

Chromatography-ICP-MS provides one way to solve the speciation problem.[12–14] The concept behind this method is that the various chemical forms in which an element might occur are separated chromatographically. Detection of the elution stream from the chromatograph is performed in an element-specific manner using ICP-MS. Much of the work on technique has come from the field of environmental analysis.[15] Some of the work in environmental analysis could be used as a basis for clinical applications, with the caveat that the matrices of clinical samples are likely to be very different from matrices in environmental samples, which may, therefore, require significant changes in the methods.

To summarize, from a technical perspective, ICP-MS is probably the premier elemental analysis method for the analysis of most elements. It is applicable to a wide range of elements. It is relatively free from interferences, and techniques have been developed to address most interferences. Its chief drawbacks are high capital cost and a high requirement for technical skill by the analyst. However, both of these drawbacks can be managed, at least in the larger reference laboratories. In this respect it differs from highly specialized techniques, such as neutron activation analysis, which require highly specialized laboratories. A second possible drawback is the complementary aspect of its strength, namely that it is an elemental analysis tool and, therefore, not well-adapted to determining the chemical form in which an element exists. This drawback can be addressed by combining chromatography with ICP-MS, thus providing speciation information in addition to elemental analysis.[13]

THERAPEUTIC DRUG MANAGEMENT

Therapeutic drug management (TDM), also known as therapeutic drug monitoring (TDM), is employed by physicians in order to select a drug regimen which will optimize therapy. It is also utilized to monitor the drug concentrations in patients to assess compliance, therapeutic efficacy, drug-drug interactions, or toxicity. The field of TDM has expanded as analytical testing for drugs has become readily available.

Therapeutic drug management can provide pharmacokinetic information, such as drug disposition, on a patient at the time of specimen collection. Pharmacokinetics describes how the body acts on the drug to absorb the drug into the blood stream, distribute the drug in the body, metabolize the drug for pharmacological activity or inactivity, and how the drug is eliminated from the body. Clinical pharmacokinetics can also be used to predict an optimal response to the drug. The dosing regimen can be changed in order to achieve therapeutic drug concentrations.

Therapeutic drug management measurements are typically from pre-dose (trough) collections, which occur immediately before the scheduled dose, once the patient has achieved steady-state concentration. Steady state occurs when the concentration of a drug in the body is in equilibrium with the rate of dose administered and the rate of elimination. Pre-dose sampling can minimize interpatient variability in pharmacokinetics in order to provide a valid comparison of a single serum/plasma concentration to the therapeutic range of a population. A random draw collection may also be useful for drugs with long half-lives or drugs that have a long distribution phase. In order for TDM to be effective, the patient's specimen should be collected at the appropriate time to determine the drug concentration in the specimen.

Therapeutic drug monitoring for free drug concentrations is performed for drugs that are highly bound to protein (typically >80% bound). Changes in the free fraction that are clinically significant can be undetected if the total concentration (protein-bound drug plus the free fraction) is monitored. Most drugs bind to albumin (drugs with acidic pKa) or α_1-acid glycoprotein (drugs with basic pKa). Alterations in plasma protein concentration can affect the balance between free and protein-bound drug. Only a free drug can cross cellular membranes and bind receptors.

Therapeutic drug management is clinically useful when there is an analytical method for the drug of interest, for drugs that have a narrow therapeutic window, and for drugs that exhibit a good correlation between the drug concentration in blood/serum and clinical effect. Therapeutic drug management can be performed on a variety of specimens; however, serum/plasma is the preferred specimen for analysis. Selection of anticoagulants for specimen collection, as well as temperature for specimen collection and storage, must be taken into consideration for specimen stability. Gel separator tubes are not recommended for specimen collection because of the potential for drug absorption into the gel.[16]

Immunoassays and chromatographic methods are commonly used in TDM. Immunoassay methods are easy to use, require minimal sample preparation and volume, are highly sensitive, and provide rapid turnaround time to result. The disadvantage is specificity due to cross-reactivity with metabolites or other drugs that are structurally similar. Chromatographic methods can be adapted to a wide variety of compounds and specimen types (blood, urine, meconium, saliva, and so forth). "Home-brew" chromatographic assays, including chromatographic methods coupled to mass spectrometry or tandem mass spectrometry, can be developed if commercial reagents are not available for TDM by immunoassay. Due to the high specificity and near-universal detection feature of LC-MS/MS, multi-analyte panels can be used to quantify multiple drugs in a single run, with minimal impact from interferences. However, the disadvantage of LC-MS/MS is that the capital purchase is expensive, and sample preparation may be more extensive and require more specimen volume than immunoassay methods. In order to minimize sample matrix affects on analysis, analytes must be ionized and separated by chromatography for detection and analysis, so the sample throughput may be limited by analysis time, and instrument operation requires higher technical skills than immunoassay. All these factors tend to limit LC-MS/MS to medium size or larger laboratories and they also make the technique poorly adapted for stat testing.

The following section for therapeutic drug monitoring will discuss the utilization of mass spectrometry in the field of TDM for the following drug classes:

immunosuppressant drugs, drugs utilized in pain management, drugs used in psychiatry, and antiepileptic drugs.

Immunosuppressants

Immunosuppressants drugs (cyclosporine A, sirolimus, tacrolimus, everolimus, mycophenolate mofetil) are used to treat autoimmune disease, allergy, multiple myeloma, chronic nephritis, and are used in organ transplantation because they suppress the immune system. Therapeutic ranges and toxic thresholds are established and used to optimize dosing of these drugs. Therapeutic drug management is important for optimizing immunosuppressant therapy due to severe consequences of underdosing (eg, graft rejection) and overdosing (eg, risk of opportunistic infections). Therapeutic drug management can also prevent drug-related toxicity (eg, kidney damage) and evaluate compliance. Mass spectrometry methods (LC-MS/MS) are commonly used to create multi-analyte panels, and to increase sensitivity and specificity compared to immunoassay methods.[17–20] Several immunoassay methods can cross react with inactive metabolites to produce results that are 20 to 60% higher than those obtained by chromatographic techniques such as HPLC or LC-MS/MS.[21–24] Immunoassay methodologies that utilize polyclonal antibodies have about 30% cross-reactivity with several metabolites and lead to falsely elevated concentrations in some patients. Current immunoassays based on monoclonal antibodies provide better specificity; however, they may not be as specific as methodologies based on HPLC or LC-MS/MS. Cyclosporine, tacrolimus, sirolimus, and everolimus accumulates in erythrocytes, thus TDM is best performed with pre-dose (trough) whole blood specimens. To release the drugs for analysis by LC-MS/MS, a step to lyse erythrocytes and precipitate proteins is typically included as part of the sample preparation procedure.[25,26]

Mycophenolate mofetil (MPA) is a reversible and uncompetitive inhibitor of inosine monophosphate dehydrogenase (IMPDH). The specimen of choice for TDM of MPA is plasma or serum. The primary metabolite of MPA is pharmacologically inactive,[27] MPA-glucuronide (MPAG). MPAG is primarily cleared by the kidneys and can accumulate and potentially cause toxicity in patients with pour kidney function.[28] MPA is extensively bound to albumin, and free MPA is monitored in patients with hypoalbuminemia, hyperbilirubinemia, and poor kidney function. TDM for MPA can be supported by immunoassay,[29] but MPA results may be overestimated by cross-reactivity with the pro-drug mycophenolate mofetil and the active metabolite, acyl-mycophenolic acide glucuronide.[30] MPA is also monitored by chromatographic methods,[31] which is the preferred methodology.

Despite the increasing number of immunoassay methods for immunosuppressant drugs, LC-MS/MS is still the method of choice. First, the consumable costs for chromatographic methods are low compared to immunoassay consumable costs. Second, chromatographic methods, particularly when coupled with tandem mass spectrometry, provide high quality results by minimizing interferences, being highly specific, and providing excellent sensitivity. Finally, multi-analyte methods using LC-MS/MS have been developed that can target several immunosuppressants in a single sample using a single method, a significant advantage for managing patients on combination immunosuppressive therapy.

Pain Management

Pharmacological management of pain involves a wide variety of drugs, including antiepileptics, tricyclic antidepressants, muscle relaxants, benzodiazepines, anesthetics, and opioids. Therapeutic drug management for pain management has become the standard practice of care and is utilized to evaluate compliance with prescribed medications and to detect noncompliance with the use of nonprescribed medications or illicit drugs.[32] Detection of patient compliance with opioid therapy is important due to concerns of misuse, risk of drug diversion, risk of drug abuse, and tolerance.[32,33] Nonprescribed medications of particular interest include other opioids or benzodiazepines, and classical drugs of abuse (eg, marijuana, cocaine, amphetamines). Compliance monitoring is usually performed on random urine specimens. However, serum or plasma specimens may be collected in patients that cannot provide a urine specimen (eg, dialysis patients).

Opioid drugs are widely used to manage chronic pain and include opiates, which are drugs derived naturally from the opium poppy plant (codeine, morphine), semisynthetic opiates (hydromorphone, hydrocodone, oxycodone, oxymorphone, heroin), and fully synthetic opioids (fentanyl, methadone, tramadol, propoxyphene, buprenorphine, meperidine). Drug screens are usually performed by immunoassay methods, and this methodology is highly susceptible to false positive/negative results, due to antibody cross-reactivity. In addition, specimen adulteration can affect antibody-antigen interaction and cause false negative or false positive results.[34] Several "opiate" immunoassay methods are limited in the number of opioids that can be detected because the antibodies are specific to detect morphine and may not cross-react with semisynthetic or synthetic opioids. In addition, these assays may lack analytical sensitivity when testing serum or plasma specimens. Laboratories that provide "pain management" testing must utilize methods that provide optimal sensitivity and specificity. Mass spectrometry methods are used for confirmation testing and provide the means for multi-analyte test panels that consist of prescription and illicit drugs. The high specificity of tests based on mass spectrometry is of particular importance in this area because incorrect results, whether false positive or false from a pain management program. The high quality of tests based on mass spectrometry reduces the risk of incorrectly classifying patient compliance.

Gas chromatography-mass spectrometry is a classical method of confirmation for drugs-of-abuse testing. However, GC-MS is rapidly being supplanted by LC-MS/MS for confirmation in many laboratories, as run times tend to be much shorter for LC-MS/MS, and because LC-MS/MS methods typically require less sample preparation effort than most GC-MS methods.

In addition to its use as a confirmation technique, mass spectrometry can also be used as a means for drug screening. Time-of-flight mass spectrometry (TOF/MS) has the ability to provide very fast full-scan data acquisition with high speed and sensitivity, and reduces the false positive/negative when compared to immunoassay drug screening.

Drugs Used in Psychiatry

Therapeutic drug management is widely used in the field of psychiatry to monitor patient compliance, to assess therapeutic efficacy, to identify slow or rapid drug metabolizers, and to monitor for toxicity. Treatment for psychological conditions is long-term and utilizes poly-drug treatment regimens, which may increase the risk of drug-drug interactions. Patient compliance is typically lower in patients with mental disorders than those with non-psychological illnesses,[35] yet treatment is also associated with high rates for suicide and mortality, thus creating a need for monitoring. Tricyclic antidepressant (TCA) drugs (eg, amitriptyline, clomipramine, imipramine) are frequently monitored in the emergency department by immunoassay. Conversely, immunoassay methods may not be available for newer generation antidepressant drugs (venlafaxine, buproprion, and so forth). Therapeutic drug management is performed on serum/plasma specimens; however, urine specimens can also be used to monitor patient compliance. Many TCA drugs have active metabolites, thus antibody cross-reactivity with active metabolites may cause an elevated result for the parent compound. High performance liquid chromatography (HPLC) methods are well established for TCA testing; however, this methodology is prone to intereferences due to multidrug therapy. Mass spectrometry methods are frequently used to create multi-analyte panels for the different classes of neuroleptic drugs (tricyclic antidepressants, selective serotonin reuptake inhibitors, serotonin and norepinephrine reuptake inhibitors, selective norepinephrine reuptake inhibitors, and antipsychotics). Mass spectrometry provides high specificity and minimizes interferences, thus it is considered the method of choice for neuroleptic drugs.[36,37]

One potential disadvantage of mass spectrometry compared to immunoassay methods is that mass spectrometry is currently not available on a stat basis for emergency care, such as for detecting tricyclic antidepressants in the emergency department. This is not a fundamental limitation of the technology but, for practical reasons (cost, staffing requirements, space requirements, potentially inefficient utilization of capital, relative newness of the technology in this sector, and other such factors), mass spectrometry is not generally available for emergency use. It will be interesting to see if mass spectrometry eventually finds its way into emergency medicine.

Antiepileptic Drugs

Antiepileptic drugs are used to prevent or attenuate seizures by altering inhibitory processes in the brain through γ-aminobutyric acid (GABA) transmission, or by inhibiting the excitatory processes through glutamate neurotransmission. Traditional antiepileptic drugs include: carbamazepine, phenytoin, phenobarbital, primidone, clonazepam, diazepam, ethosuximide, fosphenytoin, and valproic acid. Newer antiepileptic drugs consist of felbamate, gabapentin, leviteracetam, lamotrigine, oxcarbazepine, pregabalin, tiagabine, topiramate, vigabatrin, zonisamide, lacosamide, and rufinamide. These drugs can also be used to treat neuropathic pain, migraine headaches, and other psychiatric conditions.[38] Therapeutic drug management for antiepileptics is useful to manage dose adjustments and drug-drug interactions, evaluate compliance, and to assess toxicity once steady-state concentrations in the blood are achieved.[39] Antiepileptic drugs can be monitored using immunoassay methods. These assays are used commonly in the laboratory for their speed and cost. Chromatographic methods can be used to monitor parent drug and metabolites if immunoassay methods are not available. Chromatographic methods offer higher sensitivity and specificity and can allow multi-analysis of drugs to monitor polypharmacy.[40,41] Additional specificity and sample throughput would likely be obtained by converting HPLC methods to LC-MS/MS methods.

SUMMARY

Clinical toxicology can be roughly thought of as the evaluation of xenobiotic (exogenous) substances in patients. Because mass spectrometry, in addition to being highly sensitive, is both a near-universal detector and a highly specific detector, it possesses many desirable features for applications in clinical toxicology. Therefore, mass spectrometry has long been used in the clinical toxicology laboratory, such as the use of GC-MS for confirmation in drugs-of-abuse testing. The development and more recent adoption of techniques such as ICP-MS and LC-MS/MS have further expanded the role of mass spectrometry in the clinical toxicology laboratory, and this technology is currently in a rapid growth phase.

The analysis of organic drugs and toxins will likely constitute the larger number of applications of mass spectrometry in clinical toxicology. However, trace element analysis will remain an important part of clinical toxicology. Even as early as 2002 in a review article by Patriarci and colleagues, it was noted that the number of clinical applications for elemental analysis by ICP-MS were continually and rapidly increasing.[10] Clinical ICP-MS has matured to the point where, for example, whole blood lead testing has become one of the more frequently ordered tests at our laboratory. One current frontier in clinical ICP-MS is the development of methods using chromatographic separation for the purpose of speciation to distinguish highly toxic forms from less toxic forms of an element.

Therapeutic drug monitoring is an important and growing part of clinical toxicology, and the applications of mass spectrometry in this field can be seen in many ways as representative of applications of mass spectrometry in clinical toxicology as a whole. Requirements of a test for therapeutic drug monitoring vary depending on both the specific test and the circumstances surrounding the test. For example, the required turnaround time for a test for an immunosuppressant may be shorter for patients who have recently received a transplant and who are, therefore, in a dynamically changing state compared to someone on long-term treatment whose TDM needs are primarily to monitor more gradual changes in status. This somewhat mirrors the requirements for clinical drugs-of-abuse testing, where some patients (newborns undergoing drug testing in meconium samples, or patients in an emergency care setting) may require short turnaround time, whereas a longer turnaround time may be sufficient for patients who are being monitored on a long-term basis primarily to evaluate patient compliance.

Fortunately, a mass spectrometry lab can be set up to handle either situation. Our experience may be a good case in point. As a reference laboratory of national scope, most of our samples are sent to our laboratory from distant clients. While turnaround time is always a significant factor in patient testing, samples arriving from distant clients are generally not required to have very fast turnaround time (eg, < 24 hours). However, we also serve a local academic medical center. That center has a regional mission. Many of the patients at the center are visiting from out of town, or even out of state, and many of these patients require same-day results so their immunosuppressant medications can be adjusted before they return to their homes. For this particular locally based client, we are able to adjust our workflow to supply same-day turnaround for immunosuppressant analysis, provided we receive the samples prior to a certain hour of the day.

The main limitation for turnaround time in a clinical mass spectrometry lab is not the mass spectrometer itself; a mass spectrometer can typically acquire a mass spectrum within a few seconds or less. However, because clinical mass spectrometers are usually coupled to liquid chromatographs, the time to acquire the data for a single sample becomes measured in minutes rather than seconds. Sample preparation adds additional time, as does the fact that samples are usually grouped into batches rather than being run individually. Additionally, several calibrators and controls are included with each batch, and the data for a batch is normally not processed until the complete batch has been run on the instrument. Consequently, taken as a whole process, the best-case turnaround time is typically measured in hours rather than seconds.

This is not a fundamental limitation. It would be possible to operate a mass spectrometry lab with turnaround time measured in minutes rather than hours. Such a lab might even be capable of running stat samples, and one might hope to see this in the future. However, barring some breakthrough, this form of operation would not make efficient use of capital or manpower resources when compared to current practice.

In summary, mass spectrometry is rapidly expanding its role in clinical chemistry because it is a methodology that is highly specific, less prone to interferences than most other analytical techniques, and provides multi-analyte detection in a single run. This methodology is not free from limitations; matrix effect can impact the limit of detection along with ion suppression. To date, the growth in the number of mass spectrometry methods has primarily been in the area of small molecule analysis. Small molecule analysis applications can be grouped roughly into the analysis of endogenous compounds (such as certain endogenous metabolites and hormones) and the analysis of exogenous compounds, with the clinical toxicology lab focusing on the latter. The applications of mass spectrometry in clinical toxicology are rapidly proliferating, with TDM typifying this growth. The use of mass spectrometry methods is also expanding to other drug classes for TDM, which include antimicrobials, antiviral/antiretroviral drugs, and cardiovascular agents.

REFERENCES

1. Kozlovski V, Brusov V, Sulimenkov I, et al. Novel experimental arrangement developed for direct fullerene analysis by electrospray time-of-flight mass spectrometry. Rapid Commun Mass Spectrom 2004;18:780–6.
2. Meikle AW, Findling J, Kushnir MM, et al. Pseudo-Cushing syndrome caused by fenofibrate interference with urinary cortisol assayed by high-performance liquid chromatography. J Clin Endocrinol Metab 2003;88:3521–4.
3. Kushnir MM, Rockwood AL, Nelson GJ, et al. Assessing analytical specificity in quantitative analysis using tandem mass spectrometry. Clin Biochem 2005;38:319–27.
4. Available at: http://www.health.state.ny.us/publications/2584/. Accessed July 6, 2011.
5. Available at: http://www.cdc.gov/nceh/lead/casemanagement/casemanage_chap3.htm. Accessed July 6, 2011.
6. Rockwood AL, Bakowska E. Trace elements. In: Bishop ML, Fody E, Schoeff L, editors. Clinical Chemistry Techniques, Principles, and Correlations. 6th edition. Baltimore (MD): Lippincott Williams & Wilkins; 2010.
7. Available at: http://www.inorganicventures.com/tech/trace-analysis/icp-ms-measurement. Accessed July 6, 2011.
8. May T, Wiedmeyer R. A table of polyatomic interferences in ICP-MS. At Spectrosc 1998;19:150–5.
9. Marchante-Gayon J, Muniz C, Alonso J, et al. Analysis of biological materials by double focusing-inductively coupled plasma-mass spectrometry (DF-ICP-MS), vol 7. Amsterdam (The Netherlands): Elsevier Science B.V.; 2002.
10. Patriarca M, Rossi B, Menditto A. Use of atomic spectrometry (ICP-MS) in the clinical laboratory. Adv At Spectrosc 2002;7:1–51.
11. Sharp BL, Bashammakh AS, Thung CM, et al. Bayesian analysis of inductively coupled plasma mass spectra in the range 46–88 Daltons derived from biological materials. J Anal At Spectrom 2002;17:459–68.
12. Morton J, Leese E. Arsenic speciation in clinical samples: urine analysis using fast micro-liquid chromatography ICP-MS. Anal Bioanal Chem 2011;399:1781–8.
13. Heitland P, Koster HD. Comparison of different medical cases in urinary arsenic speciation by fast HPLC-ICP-MS. Int J Hyg Environ Health 2009;212:432–8.

14. Available at: http://new.americanlaboratory.com/913-Technical-Articles/596-Mercury-and-Arsenic-Speciation-Analysis-by-IC-ICP-MS/?adpi=1. Accessed July 6, 2011.

15. Popp M, Hann S, Koellensperger G. Environmental application of elemental speciation analysis based on liquid or gas chromatography hyphenated to inductively coupled plasma mass spectrometry—A review. Analytica Chimica Acta 2010;668: 114–29.

16. Dasgupta A, Dean R, Saldana S, et al. Absorption of therapeutic drugs by barrier gels in serum separator blood collection tubes. Am J Clin Path 1994;101:456–61.

17. Koal T, Deters M, Casetta B, et al. Simultaneous determination of four immunosuppressants by means of high speed and robust on-line solid phase extraction-high performance liquid chromatography-tandem mass spectrometry. J Chromatogr B Analyt Technol Biomed Life Sci 2004;805:215–22.

18. Streit F, Armstrong VW, Oellerich M. Rapid liquid chromatography-tandem mass spectrometry routine method for simultaneous determination of sirolimus, everolimus, tacrolimus, and cyclosporin A in whole blood. Clin Chem 2002;48:955–8.

19. Taylor PJ. Therapeutic drug monitoring of immunosuppressant drugs by high-performance liquid chromatography-mass spectrometry. Ther Drug Monit 2004;26:215–9.

20. Yang Z, Peng Y, Want S. Immunosuppressants: Pharmacokinetics, methods of monitoring and role of high performance liquid chromatography/mass spectrometry. Clin Appl Immunol Rev 2005;5:405–30.

21. Johnston A, Holt DW. Therapeutic drug monitoring of immunosuppressant drugs. Br J Clin Pharmacol 1999;47:339–50.

22. Brown NW, Franklin ME, Einarsdottir EN, et al. An investigation into the bias between liquid chromatography-tandem mass spectrometry and an enzyme multiplied immunoassay technique for the measurement of mycophenolic acid. Ther Drug Monit 2010;32:420–6.

23. Wallemacq P, Goffinet JS, O'Morchoe S, et al. Maine Multi-site analytical evaluation of the Abbott ARCHITECT tacrolimus assay. Ther Drug Monit 2009;31:198–204.

24. Wallenmacq P, Maine GT, Berg K, et al. Multisite analytical evaluation of the Abbott ARCHITECT cyclosporine assay. Ther Drug Monit 2010;32:145–51.

25. Wang S, Miller A. A rapid liquid chromatography-tandem mass spectrometry analysis of whole blood sirolimus using turbulent flow technology for online extraction. Clin Chem Lab Med 2008;46:1631–4.

26. Kirchner GI, Vidal C, Jacobsen W, et al. Simultaneous on-line extraction and analysis of sirolimus (rapamycin) and ciclosporin in blood by liquid chromatography-electrospray mass spectrometry. J Chromatogr B Biomed Sci Appl 1999;721:285–94.

27. Korecka M, Nikolic D, van Breemen RB, et al. The apparent inhibition of inosine monophosphate dehydrogenase by mycophenolic acid glucuronide is attributable to the presence of trace quantities of mycophenolic acid. Clin Chem 1999;45:1047–50.

28. Shipkova M, Armstrong VW, Kiehl MG, et al. Quantification of mycophenolic acid in plasma samples collected during and immediately after intravenous administration of mycophenolate mofetil. Clin Chem 2001;47:1485–8.

29. Rebollo N, Calvo MV, Martin-Suarez A, et al. Modification of the EMIT immunoassay for the measurement of unbound mycophenolic acid in plasma. Clin Biochem 2011; 44:260–3.

30. Shipkova M, Schutz E, Besenthal I, et al. Investigation of the crossreactivity of mycophenolic acid glucuronide metabolites and of mycophenolate mofetil in the Cedia MPA assay. Ther Drug Monit 2010;32:79–85.

31. Delavenne X, Juthier L, Pons B, et al. UPLC MS/MS method for quantification of mycophenolic acid and metabolites in human plasma: Application to pharmacokinetic study. Clin Chim Acta 2011;412:59–65.

32. Trescot AM, Helm S, Hansen H, et al. Opioids in the management of chronic non-cancer pain: an update of American Society of the Interventional Pain Physicians' (ASIPP) Guidelines. Pain Physician 2008;11:S5–S62.

33. Chou R, Fanciullo GJ, Fine PG, et al. Clinical guidelines for the use of chronic opioid therapy in chronic noncancer pain. J Pain 2009;10:113–30.

34. Dasgupta A. The effects of adulterants and selected ingested compounds on drugs-of-abuse testing in urine. Am J Clin Pathol 2007;128:491–503.

35. Trivedi MH, Lin EH, Katon WJ. Consensus recommendations for improving adherence, self-management, and outcomes in patients with depression. CNS Spectr 2007;12:1–27.

36. Raggi MA, Mandrioli R, Sabbioni C, et al. Atypical antipsychotics: pharmacokinetics, therapeutic drug monitoring and pharmacological interactions. Curr Med Chem 2004;11:279–96.

37. Yasui-Furukori N, Furukori H, Saito M, et al. Poor reliability of therapeutic drug monitoring data for haloperidol and bromperidol using enzyme immunoassay. Ther Drug Monit 2003;25:709–14.

38. Spina E, Perugi G. Antiepileptic drugs: indications other than epilepsy. Epileptic Disord 2004;6:57–75.

39. Johannessen SI, Landmark CJ. Value of therapeutic drug monitoring in epilepsy. Expert Rev Neurother 2008;8:929–39.

40. Kouno Y, Ishikura C, Homma M, et al. Simple and accurate high-performance liquid chromatographic method for the measurement of three antiepileptics in therapeutic drug monitoring. J Chromatogr 1993;622:47–52.

41. Juenke J, McMillin GA. Analytical Support of Classical Anticonvulsant Drug Monitoring Beyond Immunoassay: Application of Chromatographic Methods. In: Dasgupta A, editor. Advances in Chromatographic Techniques for Therapeutic Drug Monitoring. CRC Press; 2009. p. 87–103.

High Throughput Automated LC-MS/MS Analysis of Endogenous Small Molecule Biomarkers

Russell P. Grant, PhD

KEYWORDS

- Automation • Endogenous biomarkers • High throughput
- LC-MS/MS

The application of automation, liquid chromatography-tandem mass spectrometry (LC-MS/MS), and biomarker analysis in a single workflow is, at its core, a combination of numerous scientific and technical disciplines. Engineering principles of process automation, the metrological and mathematical standards of measurement error, and the chemistry of sample preparation, chromatographic separations and ion generation—together with the physics of triple quadrupoles—are optimized to provide answers to questions with a biological context. Biomarker research is clearly part of the mainstream conscience in basic research and clinical chemistry. However, one must be aware that biomarkers are the cornerstone of diagnostic medicine and certainly nothing new. In fact, the application of mass spectrometric technologies for biomarker analysis has undergone something of a revolution over the past 20 years, following the invention of atmospheric pressure ionization sources capable of analyzing liquid chromatographic eluates,[1] inducing a transition from laborious GC-MS protocols requiring thermal stability of biomarkers and derivatization.[2] Targeted analysis using atmospheric pressure ionization coupled to liquid chromatography reduced the sample preparation as well as increased chemical coverage from classical GC-MS testing.[3] Predominant drivers for mass spectrometry applications in diagnostic medicine and, more specifically, the use of liquid chromatography (LC) coupled with tandem mass spectrometry (MS/MS) for biomarkers of known clinical use are 4-fold.

First, LC-MS/MS assays do not require the generation of an antibody and are relatively easy to develop. Prerequisites are typically certified analyte standards and, preferably, a stable isotopically labeled internal standard of the biomarkers. The use

The author has no conflicts of interest and is solely employed by Laboratory Corporation of America. Research and Development, Laboratory Corporation of America, 1447 York Court, Burlington, NC 27215, USA
E-mail address: grantr@labcorp.com

Clin Lab Med 31 (2011) 429–441
doi:10.1016/j.cll.2011.07.008
0272-2712/11/$ – see front matter © 2011 Elsevier Inc. All rights reserved.

of tandem mass spectrometry with internal standardization has been demonstrated with extraordinary success for the measurement of acyl carnitines and amino acids in newborn screening.[4] In the United States, first-tier screening is performed in every state for all newborns within 3 to 5 days of birth, using semiautomated blood spot card punching, extraction, and derivatization, followed by MS/MS analysis. The known cross-reactivity of antibody-based assays and, thus, the potential for measurement error, led to a rise in applications of LC-MS/MS in the early-to-mid 2000's. Larger reference laboratories and hospitals have focused efforts towards endocrinology markers, reducing sample preparation complexity associated with radioimmunoassay (RIA) workflows while maintaining or improving result accuracy. LC-MS/MS assays for steroid hormones,[5] thyroid hormones,[6] biogenic amines,[7] targeted metabolomics,[8] and vitamins[9] are now prevalent in many laboratories. Continuing improvements in mass spectrometer sensitivity has enabled measurement of free (unbound) circulating concentrations of many of these markers.[10]

Second, the combination of isotope dilution and LC-MS/MS provides the opportunity to effect reference method levels of accuracy in a patient sample analysis. This has led to a call-to-arms for accuracy-based testing proficiency for improved concordance between clinical presentation and test results.[11–13]

Third, mass spectrometry is a mixture analysis tool. Selectivity is generated through the use of precursor and product ion pairs (transitions or selected reaction monitoring) to measure the presence of targeted analytes. These selected ions rely on intrinsic properties of molecules, namely the mass (actually mass-to-charge ratio, m/z) of the precursor ion, generally a singly protonated or deprotonated ion representative of the analyte of interest, and the mass-to-charge of a characteristic fragment of the molecule. Current triple quadrupoles can monitor hundreds of transitions per second, which equates to measuring the presence of hundreds of analytes per second as they elute from an LC column. Mixture analysis in clinical laboratories equates to multi-analyte assays, which offer significant benefits over single analyte assays, currently *de rigeur* for most immunoassay-based analysis. These benefits include full profile analysis for more than 60 amino acids and more than 75 acyl carnitines from less than 100 μL of a sample in less than 20 minutes,[14,15] and simultaneous pathway analysis for 12 steroid hormones.[16]

Last, the range of biomarkers that may be measured, the potential combinations of said biomarkers, and the increasing need for appropriate measurement leads to a simple conclusion: the application space for LC-MS/MS and mass spectrometry in general, is close to unlimited.[17]

With these favorable opportunities and obvious benefits in mind, one may ask a straightforward question, "Why aren't LC-MS/MS assays universally in place for small molecule biomarker quantitation?" The answer to that question is simple: complexity, both in terms of hardware and the analytical challenge of endogenous biomarker analysis.

The LC-MS/MS hardware currently utilized for endogenous biomarker analysis is based upon research grade instrumentation, commonly found in the pharmaceutical drug development setting. Success in implementation of LC-MS/MS has required highly skilled scientists, hence the somewhat limited uptake.[18] That being said, key applications for immunosuppressant drug monitoring[19] and the unprecedented growth in vitamin D testing have provided impetus for both manufacturers and, more importantly, financial administrators in hospital settings, to enable utility of these technologies in the clinical setting.

Many classes of clinically proven biomarkers contain structural homogeneity and orders of magnitude variance in circulating concentrations. Steroid hormone analysis represents a particular challenge, even for LC-MS/MS, and will be used as an example.

The endogenous circulating concentrations of steroid hormones can vary between greater than10 μg/mL for 17-hydroxyprogesterone, down to less than 1 pg/mL for Estradiol. Assays requiring low pg/mL levels of detection to determine insufficiency or small changes in intrinsically low levels of biomarkers are now the norm. For example, measurement of estrogenic and androgenic steroids is important in cancer research, hypogonadism, polycystic ovarian syndrome, and precocious puberty.[20,21] In contrast, analysis of xenobiotics (exogenous drugs) often requires analytical measurement in the low ng/mL range; drug metabolites are discriminated by simple LC separations or through appropriate selection of MS/MS parameters. Xenobiotics are usually structurally unique relative to circulating biomarkers, thus specificity is readily attained. The challenge of ensuring structural specificity in measurement of endogenous biomarkers represents perhaps the greatest challenge for LC-MS/MS, particularly for steroids.[22] Steroid hormones have a common 4-ring structural feature, with addition of hydroxyls, ketones, or further phase 2 metabolites (sulphates). These common features lead to a number of potentially confounding challenges. First, many steroids are geometrical isomers and, typically, a triple quadrupole cannot discriminate between them. For example, 17-hydroxyprogesterone and deoxycorticosterone are indistinguishable in a triple quadrupole mass spectrometer, and circulating concentrations can vary by greater than 5000 fold between them. Second, these large discrepancies in circulating concentrations lead to contributions from naturally occurring ^{13}C isotopes of steroids. An example of this phenomenon can be observed through contributions of mid-to-late cycle concentrations of progesterone in females, which may circulate in a greater than 100-fold excess relative to pregnenolone, and inaccurately contribute to measurement of pregnenolone without appropriate chromatographic separation. Lastly, endogenous biomarkers are often polar, requiring creativity in LC separation using hydrophobic interaction liquid chromatography,[23] or neutral, requiring additional effort to create appropriate interface conditions for precursor ion generation.[16] While these challenges may look insurmountable, a number of highly skilled researchers have published solutions to enable confidence in measurement accuracy through transition ratio monitoring,[24] structurally targeted extraction,[25] and 2-dimensional LC (2D-LC) separations.[26]

An excellent reference article was recently published, describing the broader challenges for implementation of automation and LC-MS/MS in clinical diagnostics.[27] In particular, the authors coined the phrase "heterogeneous instrument configurations" to describe open architecture platforms (liquid handlers, LC, and MS technology). Interested readers should review this article for a more thorough review of published literature. Due to the limited literature in this area, this article will describe published highlights in technology and application, together with our own experiences and successes in automated sample preparation, high throughput LC-MS/MS, and automated data reduction. The article will close with emerging technologies poised for clinical utility.

AUTOMATED SAMPLE PREPARATION USING LIQUID HANDLING TECHNOLOGY

The majority of automated sample preparation processes employed in endogenous biomarker quantitation have arisen from pharmaceutical drug development applications, namely batch mode of operation, in which a number of samples are prepared simultaneously using a common protocol. This is clearly at odds with the need for real-time sample processing, using an automated line for clinical diagnostics. A limited number of publications have demonstrated the "potential" to execute much of the pre-analytical and analytical processes in real-time mode with minimal human intervention.[28,29] These technologies center on solid-phase extraction (SPE)-based sample enrichment, often using dual arm or dual channel pipetting systems with integrated LC-MS/MS platforms. Current practice almost universally emulates batch process mode, but not without challenges. No vendor currently provides liquid handling technology that can be considered "plug and play" for integration with LC-MS/MS.[27] In practice, our experience suggests at least 3 to 6 months of dedicated and highly competent full time equivalent (FTE) resource is required for initial functional execution of these technologies to ensure accurate and precise handling of the multitude of matrices tested in the clinical setting. This inherent gap in capability is not insurmountable; in fact, similar process control was developed for current state-of-the-art autoanalyzers. Unfortunately, the diagnostic industry and liquid handling/mass spectrometry manufacturers represent 2 discreet camps, with different agendas.

These barriers have not dissuaded laboratories from implementing flexible (islands of automation) systems for partial automation of many processes.[30] The goals, quality enhancements, and process improvements are described as follows. For real-time process, liquid-handling technologies must both read the barcode of a specimen and determine which tests are appropriate (LIMS integration), as many preparative schemes are biomarker- or class-specific. To our knowledge, these fully integrated systems have not yet been published for LC-MS/MS assays. In our laboratory, the first step of offline liquid handling systems is to create an import list (sample order) of calibrators, quality controls, and samples prior to or during pipetting through positive barcode ID scanning. Practically, samples are presorted for a specific assay prior to this step. The import list serves 2 main functions: the ability to initiate in-process sample tracking (providence) of batch content and sample order upload to LC-MS/MS systems. The next 2 steps are universal for all quantitative mass spectrometry assays: accurate and precise pipetting of batch content, and addition of stable labeled internal standards into a 96-well plate, sample vial, or process tube.

These first 3 steps may be the sum total of the offline automation requirement for some assays. The internal standard is usually delivered in a miscible diluent for direct injection onto a 2D LC system, where enrichment and purification occurs in the first dimension,[31,32] or using a precipitating reagent to remove protein content that can reduce LC column lifetime for direct injection of the supernatant.[23] Both techniques involve dilution of the sample, and are generally only utilized when circulating concentrations of biomarkers are such that current LC-MS/MS systems have the requisite sensitivity. Sample vial or 96-well plate sealing, mixing, and centrifugation are routinely performed offline prior to LC-MS/MS analysis. More commonly, some degree of selectivity is incorporated during the offline sample preparation process through solid phase extraction, liquid-liquid extraction, or antibody capture. Inclusion of on-deck vacuum devices facilitates additional automated workflows, such as ultrafiltration[33] and solid phase extraction.[34] These workflows predominantly use 96-well preparation devices, with final purified and enriched biomarker elution into a fresh 96-well plate for subsequent off-deck handling prior to analysis. High through-put LC-MS/MS analysis requires robustness to be built into sample preparation to reduce system downtime. Liquid-liquid extraction (LLE) is often a desireable approach, particularly when concentrating samples.[35] The automation of LLE represents a particularly challenging application for automation because of the volatility of nonpolar solvents used for steroid analysis.[36] Ratios of 10:1 solvent to sample are often used for LLE precluding the use of 96-well plates when 200 to 500 μL of sample is required for sensitive analyte detection, as is often the case for underivitized analysis of endogenous steroid hormones. Recent advances in 96-well plate technologies for ease of handling involving freezing the sample plate, inverting, and pouring the solvent extract into a clean plate have been published.[37] In our laboratory,

Tecan EVO
(8 Tip)

Tube-plate (PMP) – 23040 samples/day (240)
Tube-plate/Dilute/PPT – 11520 samples/day (120)
Plate-plate - 11520 samples/day (120)

Hamilton
Star/Starlet
(16-Tip)

Tube-plate (TADM) – 34560 samples/day (360)
Tube-plate and Dilute/PPT – 17280 samples/day (180)
Tube-plate and LLE – 8640 samples/day (90)
Plate-plate-23040 samples/day (240)

Beckman
Biomek FX
(96-Tip)

Plate - plate – 69120 samples/day (720)
Diluent/PPT in plate – 138240 samples/day (1440)

Fig. 1. Examples of liquid handling technologies employed within our laboratories for plate formatting (parent tube to 96-well plate) and plate to plate processing. Representative samples per day and plates per day capacities we have achieved are denoted for specific automated processes.

we have implemented 96-well plate supported liquid extraction (SLE), a hybrid of SPE and LLE, allowing highly successful industrialization of the liquid-liquid extraction process.[38]

Current state-of-the art offline liquid handling systems are designed with the avoidance of gross handling errors in mind. Most liquid handlers employ some means of determining liquid levels, sample clots, and air bubbles. However, both Hamilton (total aspiration and dispensing monitoring, TADM, Hamilton Robotics, Reno NV) and Tecan (pressure monitored pipetting, PMP, Tecan Group Ltd. Switzerland) employ technologies that supercede those found in autoanalyzers. Both TADM and PMP monitor "pressure curves", essentially a functional real-time monitoring during sample aspiration and dispensing that enables rapid review of sample handling post preparation where these steps are in question. We have universally employed these flexible pipetting platforms for our LC-MS/MS assays. Our practical experiences regarding these liquid handling systems with disposable tips are shown in **Fig. 1**. Practically realized per sample and per 96-well plate capacities over 24 hours are shown, based upon batch sizes of 192 samples (two 96-well plates per assay batch). When using these technologies for parent tube to 96-well plate sample plus internal standard addition, 240 to 360 of the 96-well plates per day throughput has been realized. When performing additional steps such as precipitating and mixing on deck or transferring from plate-to-plate, the capacities of these systems are reduced by a factor of 2. The primary throughput differences between these platforms relates to the number of pipetting tips (16 tips for the Hamilton Star and 8 tips for the Tecan Evo). Additionally, the Hamilton Star provides the ability to perform highly volatile LLE extract aspiration through micro-stepping of the dilutor, equalizing the pressure buildup during solvent transport from an intermediate tube or plate to a fresh tube or 96-well plate. Following accurate addition of sample and internal standards (plate formatting), additional sample processing efficiency is gained through the use of 96-tip pipetting systems (plate processing). Additionally, it is important to note that the process requirements for 96-tip liquid handling center around precision across tips and speed of plate-to-plate processing, not accuracy. We employ a 3:1 ratio of plate formatting (flexible 8/16 tip liquid handlers) to plate processing (96-head pipetting) to optimize the total workflow through utilizing the benefits of both technologies.

A review of the plethora of LC-MS/MS assays currently employed in diagnostic medicine, and those described above, indicates the need for method-specific workflows, particularly in low-level biomarker quantitation. The value of selectivity in sample preparation cannot be understated, as accurate measurement is often generated through the additive selectivity generated through the orthogonal steps of sample preparation, LC separation, and MS/MS detection.[36]

HIGH THROUGHPUT LIQUID CHROMATOGRAPHY-TANDEM MASS SPECTROMETRY LC-MS/MS

The definition of "high-throughput" within the context of any application is subjective. In clinical diagnostics, the definition of high-throughput is often defined by the capacity of the majority of commercial autoanalyzers, namely greater than 2000 samples per 24 hours per system. Arguments abound regarding the value of counting the discreet analytes measured in unit time as a measure of throughput for multi-analyte LC-MS/MS assays. However, these tend to misrepresent the diagnostic challenge, where sample numbers, not independent biomarkers, are the primary concern. Parallels exist between operation of LC-MS/MS and automated liquid handling; namely, batch mode of operation still dominates the application space. This is a function of the independent nature of highly specific LC separations, often an absolute requirement. With the noted exceptions of immunosuppressant analysis and other assays required for pre, mid, and post-surgical care, the vast majority of endogenous biomarker analysis is performed under less stringent turnaround time constraints (>48 hours), thus biomarker-specific assays are not a concern in most laboratories.

The technique of LC-MS/MS can be considered a truly orthogonal technique; a combination of chemistry and physics principles. In our laboratory, the acquisition parameter settings (MS/MS parameters) are only moderately flexible. Significant effort is expended in screening potential transitions for measurement of analytes, prior to selection of the final quantifying and qualifying transitions.[24] As noted previously, selectivity is generated through a combination of all techniques. In practice, lower sensitivity transitions are often chosen to provide additional confidence in measurement accuracy for endogenous biomarkers such as vitamin D, biogenic amines, and nonpolar steroid hormones.[30] While this may seem illogical at first glance, increased selectivity in detection provides the opportunity for reduced chromatographic cycle times, resulting in increased throughput for analysis. This paradigm places an unprecedented burden on the LC separation to ensure quality of results. No single LC column or modality of operation is sufficient to address the primary challenge in clinical diagnostics analyses: the unequivocal assignment of accurate and precise measurement of endogenous biomarkers from an uncontrolled population.

Laboratories employing LC-MS/MS technologies generally fall into one of 3 categories, largely driven by sample volumes. The larger laboratories tend to focus on 1 to 4 analytes per assay, continually looking to reduce cycle-time per injection through incremental improvements in each step of the process.[39,40] It should be noted that the majority of these laboratories have developed these capabilities within the steroid hormone analysis setting and have aggressive strategies focused upon obsoletion of antiquated technologies (RIA) and reducing costs per test.[16,27,30,36] The second category represents those laboratories that attempt to measure as many analytes as feasibly possible in a single sample injection (multiplexed analyte measurement). Larger laboratories have gravitated in this direction for a few applications.[41] However; the throughput per sample is often the deal breaker, such that assays requiring 10 to 60 minutes of mass spectrometer acquisition time per injection have limited scalability. Laboratories falling into the second group are often represented by research laboratories with medical university affiliations[15] or contract research organizations.[42] The common theme is to provide assays for basic and applied researchers. The last category represents the majority of hospital laboratories. Requirements for analysis of immunosuppressants and vitamin D have enabled the initial purchase of technology and expertise. However, pressure to measure more analytes and accelerate the return on investment is the norm. These laboratories tend to develop generic LC methods to enable analysis of different analytes without manual intervention; this strategy leads to more exhaustive sample preparation, often involving derivatization.

There are 2 potential solutions to increase LC-MS/MS throughput. First, one can increase the specificity of sample preparation, thereby sacrificing the fidelity of the LC separation. While this may appear initially promising, the complexity of endogenous analyte measurement often detracts from the potential benefits of this decision. Fortunately, there have been dramatic improvements in LC technologies in the last 15 years. The biggest contributor to LC-MS/MS throughput in clinical diagnostics has incorporated a concept known as staggered parallel multiplexed LC, almost exclusively effected using an ARIA system in larger laboratories.[43] This concept is described in the upper portion of **Fig. 2**. Eluent from the first channel is selected by means of a valve and diverted to a mass spectrometer for measurement of the portion of the LC chromatogram that contains the biomarkers of interest (shaded region). Intelligent timing algorithms enable the pre-injection of a sample onto the second channel and, following a fixed mass spectrometer reset delay (file closure by the mass spectrometer), the second channel is selected for detection (second shaded region). This cycle continues through the 4 channels and back to the first channel to repeat the cycle. This staggering of injections leads to the staggered parallel nomenclature. **Fig. 2** (lower section) shows chromatograms representing lower and upper limits of quantification (LLOQ and ULOQ, respectively), together with representative calibration curves using an ARIA TX4 system coupled with an API 4000 triple quadrupole mass spectrometer in our laboratory (circa 2004). Each channel functionally executes an independent steroid assay, either a single biomarker measurement (for instance, cortisol in serum or cortisol in urine or testosterone) or a multi-analyte assay (for instance, progesterone and 17-hydroxyprogesterone). Note that the pre-analytical sample preparation varies per channel, LLE = liquid extraction, turbulent flow chromatography (TFC) denotes direct injection with turbulent flow extraction, LC-LC = 2D LC separation. This data represents state-of-the art circa 2004, with an instrument capacity approaching 1000 samples per mass spectrometer per day, based on 4 to 4.5 minute LC cycle times.

Second, improvements in single channel LC technology have occurred, loosely termed ultra performance liquid chromatography.[44] These improvements have led to increased operating pressure of LC pumps, facilitating higher flow rates for LC separations, minimizing the dead volume in LC hardware, reducing the reconditioning time between injections, increasing the variety and reproducibility of LC columns, providing alternate selectivities, and reducing the chromatographic particle size, thereby improving the resolving capacity of a given LC separation. The net result is an overall reduction in inject-to-inject time, leading to increased LC-MS/MS throughput.[45] These upgrades in single channel LC systems have been adopted in the multiplexing system described in **Fig. 2**. Through 2005 and 2006, enhanced autosampler washing systems became commercially available, resulting in carry-over removal and inject-to-inject cycle time reductions from 2 minutes to 48 seconds. Liquid chromatog-

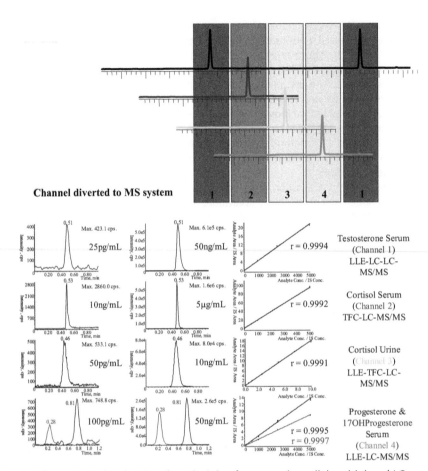

Fig. 2. Schematic (top) indicating the principle of staggered parallel multiplexed LC operation. Shaded regions denote portions of eluent that are diverted to a mass spectrometer for detection. Example chromatograms of independent steroid hormone measurements performed using staggered parallel multiplexed LC on an ARIA TX4 with an API4000 triple quadrupole MS system operating at 1000 samples/day, circa 2004. Lower (left) and upper (middle) limits of quantification are shown for independent assays, together with calibration curves, pre-analytical and chromatographic processes.

raphy pump hardware dead volumes were reduced from 850 to 120μL and operating pressures were increased from 400 to 600 bar. Further, next generation triple quadrupoles were developed with sensitivity increases greater than 20-fold. The net results of these enhancements was an increase of LC-MS/MS throughout to greater than 2500 samples per mass spectrometer per day, representing almost 2 samples per minute acquisition rates and 90% mass spectrometer detection utility.[14,23,30,36,38]

AUTOMATED CHROMATOGRAPHIC DATA REVIEW

The inevitable outcome of driving automation and chromatographic multiplexing systems is a deluge of data for review. Data generated by LC-MS/MS technologies is noisier than LC-UV due to the complex ion-molecule reactions that occur in the

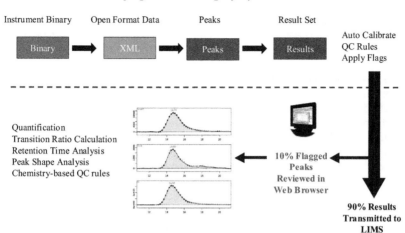

Fig. 3. Schematic workflow for Automated data reduction using Ascent™. Processes above the dotted line occur 5-10 minutes after sample/batch acquisition. (*Slide courtesy of* Dr Randall Julian, Indigo Biosciences with full permission.)

interface. Automated peak smoothing enables software tools to automatically integrate this "noisy" data; however, as perfectionists, it is easy to review a chromatogram and decide you could do a better job. The net result of such actions is negligible within the context of accuracy; the real penalty is technologist time. As an example, we have recently implemented an LC-MS/MS assay for 60+ amino acids,[14] converting the protocol from a 3-hour per sample LC-UV assay with a post-column ninhydrin derivatization. The LC-UV assay generated data at 2 wavelengths, providing 2 chromatograms to review with 40+ peaks. The new assay generates 2 chromatograms for each amino acid (quantifying and qualifying transitions), plus 40 chromatograms for the internal standards. This equates to 160 chromatograms per patient, or an 80-fold increase in data for review. This explosion in dimensionality of data requiring review has been a challenge throughout the implementation of LC-MS/MS technologies in general. In clinical diagnostics, moving from immunoassays (numbers only) and nonspecific detection strategies such as UV (1 or 2 chromatograms) has created a data review bottleneck.

Over the last 4 years, we have worked with Indigo Biosciences to correct this deficiency and help to provide a commercially available fix for this conundrum. A schematic of the basic workflow of Ascent is shown in **Fig. 3**. The process is as follows: data acquired in either real-time or batch mode is swept from the acquisition PC and converted to a common format, XML. This first step removes the need for vendor-specific data reduction systems and sets the path for a common data review software type, irrespective of source data. Chromatographic peaks are selected, integrated, and assigned a role (calibrator, quality control, blank, or unknown for example). Batch calibration is applied and calibration plots are generated, with the opportunity to automatically include/exclude calibrator points based on residual error (bias) or signal to noise criteria. Quality controls are back-calculated against the calibrators and Westgard rules are applied. Additional parameters such as transition ratio monitoring (a measure of peak purity for LC-MS/MS assays) and internal standard variability (a measure of recovery variance and instrument drift) are computed and plotted. Finally, a provisional results table is generated, with additional chromatography rules applied, such as peak asymmetry and retention time drift. Individual chromatograms (biomarker transitions) are flagged for review if they fail pre-set quality rules. These processes occur automatically and are offline in either real-time or 5 to 10 minutes after a batch has been acquired (above the dashed line in **Fig. 3**).

These steps replace the first stage of data reduction and review that, in our laboratory, is 100% of data. The inclusion of smart flags to observe for peak purity, instrument drift, recovery variance, analytical errors, and clinical abnormality aims to reduce the required review time and refocus data review on sample chromatograms that require attention. This reduces the amount of time spent reviewing results that have no flags and, thus, are considered acceptable for release. Based upon our experience, well-developed and validated assays generate less than 10% flags, predominantly for clinical flagging reasons, not for reasons of analytical error. In-house studies indicate that these processes reduce overall data review from 200% to 10% total chromatographic review time per assay. Example chromatograms for our methylmalonic acid (MMA) LC-MS/MS assay are shown in **Fig. 3** (lower lefthand side). Chromatographic peaks (from top to bottom) are the MMA quantifying transition (which is used to generate final results), the MMA quantifying ion (to assess peak purity of the quantifying transition), and the D_4-MMA stable labeled internal standard quantifying transition. We currently generate more than 2500 chromatograms daily for this assay and provisional studies have indicated a feasible reduction in review time from more than 12 hours to less than 60 minutes per day using automated data review.

EMERGING TECHNOLOGIES AND CONCLUDING REMARKS

This article has described a number of the state-of-the-art technologies that are currently employed in clinical laboratories throughout the United States. To many, these may seem unattainable, which is a real concern. Many of these concepts and applications have evolved over the past 5 years and, to those in the field, appear to be second nature. That is the benefit of being on the other side of the hard work necessary to drive these concepts to practical reality. For those readers that are looking for the next step, there are 2 additional stages of scientific evolution that clinical laboratories are evaluating that will have a profound impact in this arena. First, the fundamental capabilities of antibodies should not be ignored. As described above, pre-analytical sample preparation can be convoluted and laborious, even for high performance LC-MS/MS technologies. A number of assays are immediately amenable to coupling antibodies and mass spectrometric techniques for small molecule biomarker assays. Functionally, antibody selectivity is additive to the overall confidence in measurement, and multiple degrees of orthogonal selectivity can enable overall reductions in cycle time (internal unpublished data). Second, a recent publication[46] essentially described 3 stages of multiplexing. The automated LC-MS/MS assay described measures for both 25-hydroxyvitamin D2 and D3 (Vitamin D) in a patient sample. This is considered analyte multiplexing. Each sample is derivatized with a reagent, however, and this is the novelty; the derivatizing reagent for sample 1 is slightly different than for sample 2, all the way up to sample 5. This means that the vitamin D content in each of these 5 samples is structurally unique and may be resolved by the mass spectrometer. These 5 samples are then pooled together to generate a single composite sample for injection. This is considered sample multiplexing. Composite samples are injected using a 4-channel LC system coupled to a triple quadrupole mass spectrometer, employing the principles described above (chromatographic multiplexing). The net result of these combined stages of multiplexing is an astounding 7200 samples per system per day, truly competitive with autoanalyzer technologies in current use.

The path forward for greater implementation of mass spectrometry must be focused on a systems solution; this has to be driven by manufacturers such that the future application of these technologies does not remain concentrated in specialized laboratories. Tentative efforts have been made by manufacturers to create template LC-MS/MS methods with components for calibration, quality control, and separation. This step is potentially fraught with regulatory constraints and, thus, cannot be considered the status quo. The goal is clear: clinical LC-MS/MS systems must emulate autoanalyzer platforms in terms of simplicity, ruggedness, and throughput. Clinical mass spectrometers must index with the automated line of clinical laboratories and provide random access real-time analysis, without the need for data review.

REFERENCES

1. Proctor CJ, Todd JFJ. Atmospheric pressure ionization mass spectrometry. Org Mass Spec 1983;18;509–16.
2. Niwa T. Metabolic profiling with gas chromatography-mass spectrometry and its application to clinical medicine. J Chromatogr 1986;370;313–45.
3. Van Bocxlaer JF, Clauwaert KM, Lambert WE, et al. Liquid chromatography-mass spectrometry in forensic toxicology. Mass Spectrom Rev 2000;19;165–214.
4. Chace D. Mass spectrometry in newborn and metabolic screening: historical perspective and future directions. J Mass Spectrom 2009;44;163–70.
5. Shackleton CH. Mass spectrometry in the diagnosis of steroid related disorders and in hypertension research. J Steroid Biochem Mol Biol 1993;45:127–40.
6. Soldin OP, Soldin SJ. Thyroid hormone testing by tandem mass spectrometry. Clin Biochem 2011;44:89–94.
7. de Jong WH, de Vries EG, Kema IP. Current status and future developments of LC-MS/MS in clinical chemistry for quantification of biogenic amines. Clin Biochem 2011;44:95–103.
8. Schmedes A, Brandslund I. Analysis of methylmalonic acid in plasma by liquid chromatography-tandem mass spectrometry. Clin Chem 2006;52:754–7.
9. Chace DH. Mass spectrometry in the clinical laboratory. Chem Rev 2001;101: 445–78.
10. Yue B, Rockwood AL, Sandrock T, et al. Free thyroid hormones in serum by direct equilibrium dialysis and online solid-phase extraction–liquid chromatography/tandem mass spectrometry. Clin Chem 2008;54:642–51.
11. Bhasin S, Zhang A, Coviello A, et al. The impact of assay quality and reference ranges on clinical decision making in the diagnosis of androgen disorders. Steroids 2008;73: 1311–7.
12. Thienpont LM, Van Uytfanghe K, Blincko S, et al. State-of-the-art of serum testosterone measurement by isotope dilution-liquid chromatography-tandem mass spectrometry. Clin Chem 2008;54:1290–7.
13. Singh RJ, Grebe SK, Yue B, et al. Precisely wrong? Urinary fractionated metanephrines and peer-based laboratory proficiency testing. Clin Chem 2005;51:472–3; discussion 473–4.
14. Rappold B, Grant RP, Holland P. Quantitative underivatized amino acid analysis - development, dimensionality, data reduction and diagnostic utility. Conference abstracts and Proceedings of the 57th American Society for Mass Spectrometry, Salt Lake City, 2010, MOB 176.

15. Minkler P, Stoll MSK, Ingalls S, et al. Selective, Accurate, and Precise Quantification of Acylcarnitines in Human Urine, Plasma, and Skeletal Muscle by HPLC-MS/MS, Conference abstracts and Proceedings of the 58th American Society for Mass Spectrometry. Denver, 2011, MP349.
16. Guo T, Taylor RL, Singh RJ, et al. Simultaneous determination of 12 steroids by isotope dilution liquid chromatography-photospray ionization tandem mass spectrometry. Clin Chim Acta 2006;372:76–82.
17. Grebe SK, Singh RJ. LC-MS/MS in the Clinical Laboratory - Where to From Here?, Clin Biochem Rev 2011;32:5–31.
18. Vogeser M, Seger C. Pitfalls associated with the use of liquid chromatography-tandem mass spectrometry in the clinical laboratory. Clin Chem 2010;56:1234–44.
19. Taylor PJ. Therapeutic drug monitoring of immunosuppressant drugs by high performance liquid chromatography-mass spectrometry. Ther Drug Monit 2004;26:215–9.
20. Santen RJ, Lee JS, Wang S, et al. Potential role of ultra-sensitive estradiol assays in estimating the risk of breast cancer and fractures. Steroids 2008;73:1318–21.
21. Stanczyk FZ, Clarke NJ. Advantages and challenges of mass spectrometry assays for steroid hormones. J Steroid Biochem Mol Biol 2010;121:491–5.
22. Kushnir MM, Rockwood AL, Roberts WL, et al. Liquid chromatography tandem mass spectrometry for analysis of steroids in clinical laboratories. Clin Biochem 2011;44:77–88.
23. Rappold B, Grant RP, Crawford M, et al. An Empirical Marriage of HILIC and Clinical Diagnostics: Exploration and Expansion of Development, Utility and Doctrine, Conference abstracts and Proceedings of the 58th American Society for Mass Spectrometry. Denver, 2011, TOG125.
24. Kushnir MM, Rockwood AL, Nelson GJ. Assessing analytical specificity in quantitative analysis using tandem mass spectrometry. Clin Biochem 2005;38;319–27.
25. Kushnir MM, Urry FM, Frank EL, et al. Analysis of catecholamines in urine by positive-ion electrospray tandem mass spectrometry. Clin Chem 2002;48;323–31.
26. Kushnir MM, Rockwood AL, Bergquist J. High-sensitivity tandem mass spectrometry assay for serum estrone and estradiol. Am J Clin Pathol 2008;129;530–9.
27. Vogeser M, Kirchoff F. Progress in automation of LC-MS in laboratory medicine. Clin Biochem 2011;44:4–13.
28. Blomberg LG. Two new techniques for sample preparation in bioanalysis: microextraction in packed sorbent (MEPS) and use of a bonded monolith as sorbent for sample preparation in polypropylene tips for 96-well plates. Anal Bioanal Chem 2009;393:797–807.
29. De Jong WHA, Graham KS, van der Molen JC. Plasma free metanephrine measurement using automated online solid phase extraction HPLC-tandem mass spectrometry. Clin Chem 2007;53:1684–93.
30. Grant RP, Rappold B, Holland P. Ultra High-Throughput Quantitative LC-MS/MS in a Clinical Diagnostics Laboratory - Breaking the 2000 samples/system/day barrier., Conference abstracts and Proceedings of the 56th American Society for Mass Spectrometry. Philadelphia, 2009, MOB.
31. Rauh M, Groschl M, Rascher W, et al. Automated, fast and sensitive quantification of 17alpha-hydroxyprogesterone, androstenedione, and testosterone by tandem mass spectrometry with online extraction. Steroids 2006;71:450–8.
32. Henriksen T, Hellestrom PR, Poulsen HE, et al. Automated method for the direct analysis of 8-oxo-guanosine and 8-oxo-2'-deoxyguanosine in human urine using ultraperformance liquid chromatography and tandem mass spectrometry. Free Radic Biol Med 2009;47:629–35.

33. Blom HJ, van Rooij A, Hogeveen M. A simple high-throughput method for the determination of plasma methylmalonic acid by liquid chromatography-tandem mass spectrometry. Clin Chem Lab Med 2007;45:645–50.
34. Alvarez Sánchez B, Capote FP, Jiménez JR, et al. Automated solid-phase extraction for concentration and clean-up of female steroid hormones prior to liquid chromatography-electrospray ionization-tandem mass spectrometry: an approach to lipidomics. J Chromatogr A 2008;1207:46–54.
35. Bonfiglio R, King RC, Olah TV, et al. The effects of sample preparation methods on the variability of the electrospray ionization response for model drug compounds. Rapid Commun Mass Spectrom 1999;13:1175–85.
36. Grant RP, Rappold B, Holland P. Steroid Analysis in Clinical Diagnostics: Selectivity, Sensitivity and Speed, Abstracts and Proceedings of the 3rd Mass Spectrometry and the Clinical Laboratory. San Diego, 2011.
37. Hoofnagle AN, Laha TJ, Donaldson TF, A rubber transfer gasket to improve the throughput of liquid-liquid extraction in 96 well plates: application to vitamin D testing. J Chromogr B Analyt Technol Biomed Life Sci 2010;878:1639–42.
38. Crawford M, Rappold B, Grant RP. Clinical Testosterone Analysis for the Masses: Taking Testosterone Testing to the Next Level, Abstracts and Proceedings of the 3rd Mass Spectrometry and the Clinical Laboratory. San Diego, 2011.
39. Singh RJ. Quantitation of 17-OH-progesterone (OHPG) for diagnosis of congenital adrenal hyperplasia (CAH). Methods Mol Biol 2010;603:271–7.
40. Kushnir MM, Rockwood AL, Bergquist J. Liquid chromatography-tandem mass spectrometry applications in endocrinology. Mass Spectrom Rev 2010;29:480–502.
41. Lacey J, Casetta B, Daniels SB, et al. Amino Acid Quantitation in Plasma, Urine and CSF by iTRAQ™ Reagent Amino Acid Analysis Kit and MS/MS, Conference abstracts and Proceedings of the 56th American Society for Mass Spectrometry. Denver, 2008, 205.
42. Evans AM, DeHaven CD, Barrett T, et al. Integrated, nontargeted ultrahigh performance liquid chromatography/electrospray ionization tandem mass spectrometry platform for the identification and relative quantification of the small-molecule complement of biological systems. Anal Chem 2009;81:6656–67.
43. Singh RJ. Validation of a high throughput method for serum/plasma testosterone using liquid chromatography tandem mass spectrometry (LC-MS/MS). Steroids 2008;73:1339–44.
44. MacNair JE, Lewis KC, Jorgenson JW. Ultrahigh-pressure reversed-phase liquid chromatography in packed capillary columns. Anal Chem 1997;69:983–9.
45. Pieters S, Dejaegher B, Vander Heyden Y. Emerging analytical separation techniques with high throughput potential for pharmaceutical analysis, part I: Stationary phase and instrumental developments in LC. Comb Chem High Throughput Screen 2010;13:510–29. Review.
46. Netzel BC, Cradic KW, Bro ET, et al. Increasing liquid chromatography-tandem mass spectrometry throughput by mass tagging: a sample-multiplexed high-throughput assay for 25-hydroxyvitamin D2 and D3. Clin Chem 2011;57:431–40.

Regulatory Considerations for Clinical Mass Spectrometry: Multiple Reaction Monitoring

Emily S. Boja, PhD[a], Henry Rodriguez, PhD, MBA[b],*

KEYWORDS

- Clinical proteomics • Multiplex
- Multiple reaction monitoring mass spectrometry
- In vitro diagnostics • Regulatory science

Clinical proteomics undoubtedly holds great promise in medicine as it provides valuable information to the study of diseases at the molecular level, and has the potential to discover biomarkers for disease states. In cancer, the discovery of protein/peptide signatures "leaked" by tumors into clinically accessible fluids such as blood leads to the possibility of developing quantitative assays for diagnosing cancer at an early stage or monitoring response to therapy. To date, there have been over 1,200 cancer-related protein biomarker candidates published in the scientific literature.[1] However, the rate of introduction of new protein biomarkers to market as approved by the Food and Drug Administration (FDA)[1] has remained stagnant over the past 15 years, averaging 1.5 new proteins per year for all diseases.[2] This discrepancy points to an ineffective translation of proteomics from the bench to the bedside. Several factors have been identified as rate-limiting for the translation of new clinical biomarkers,[3–6] including: analytical variability within/across platforms, poor study design without proper statistical power and cohort selection, inability of credentialing biomarker candidates (preclinical) prior to costly and time-consuming clinical qualification using well-established methodologies such as the

---

The authors have nothing to disclose.

[a] Office of Cancer Clinical Proteomics Research, National Cancer Institute, National Institutes of Health, 31 Center Drive, MS 2590, Bethesda, MD 20892, USA

[b] Office of Cancer Clinical Proteomics Research, National Cancer Institute, National Institutes of Health, 31 Center Drive, MS 2590, Bethesda, MD 20892, USA

* Corresponding author.

E-mail address: rodriguezh@mail.nih.gov

[1] The FDA "approves" premarket application submissions (PMA) for Class I devices and "clears" 510(k) submissions for Class II devices. For the purposes of this article, the words "approved" and "cleared" have the same meaning and are not related to any proposed or real classification decision for any device unless they are specifically defined in the text.

Clin Lab Med 31 (2011) 443–453

doi:10.1016/j.cll.2011.07.001

0272-2712/11/$ – see front matter Published by Elsevier Inc.

labmed.theclinics.com

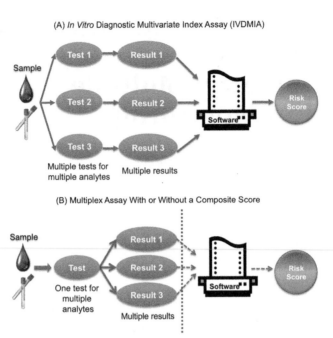

Fig. 1. Graphical Representations of IVDMIA and Multiplex Devices. (*A*) An IVDMIA example is the 3 tests representing each independently performed, multivariate measurements which ultimately provide a composite risk score calculated from software algorithm. (*B*) A multiplex assay (in this case, a triplex) simultaneously measures multiple analytes via a common test and generates several measurements (outputs) either without a risk score, which would report the results for each analyte separately and would not continue through to software (ie, stops at the dashed vertical line), or would generate a composite risk score through an algorithm (ie, goes beyond the dashed vertical line).

Enzyme-Linked Immunosorbent Assay (ELISA), and failure to validate biomarkers in clinical qualification studies. In addition, a major hurdle in translational science is the research community's lack of understanding of the evaluation criteria required by the FDA to approve protein-based assays, as proteomic studies aiming to produce useful biomarkers and their associated assays for the clinic need to be properly designed from biospecimen cohorts to data and statistical analyses on analytically validated platforms.

To remedy this problem, the National Cancer Institute's Clinical Proteomic Technologies for Cancer initiative (NCI-CPTC) proposed a pre-clinical "verification" stage to credential biomarker candidates on analytically robust platforms, thus hoping to close the gap between protein-based biomarker discovery and the FDA's clinical qualification (**Fig. 1**). Consequently, NCI-CPTC investigators and their FDA colleagues published 2 mock premarket notifications, also called PMN or 510(k) pre-submission documents, to help the scientific community interested in developing diagnostic products for the clinic understand the likely analytical evaluation requirements for multiplex protein-based tests.[7] These first-of-their-kind educational materials are important since section 510(k) of the Food, Drug and Cosmetic Act requires device manufacturers to register and notify the FDA of their intent to market a medical device at least 90 days in advance. This allows FDA to determine whether the device is equivalent to a predicate device. The 2 types of multiplex protein-based assays described in these documents include, as examples, multiple reaction monitoring mass spectrometry (MRM-MS) on a triple quadrupole mass spectrometer (TQMS),[7–10] and immunological arrays.[7,11]

FDA's CLASSIFICATION OF IVDs

A thorough review of the FDA's classification of in vitro diagnostics (IVDs) has been provided by Mansfield and colleagues.[12] Briefly, the level of device regulation required by the FDA is dependent upon a device's risk to health outcomes in patients based on the device's performance. The type of submission required depends on which class a device falls into and is highly dependent on the claim of the test (ie, intended use and indication for use) instead of the nature of the analytes. Three classification categories (classes) of medical devices have been defined by the Food, Drug and Cosmetic Act, which must be manufactured under a quality assurance program, be suitable for the claimed intended use, be adequately packaged and properly labeled, and have establishment registration and device listing forms on file with the FDA[12]:

- *Class I (low-risk).* Low-risk devices are generally classified as Class I, most of which are exempt from FDA premarket review. A few Class I devices are additionally exempt from most Good Manufacturing Practice (GMP) requirements. However, they are not exempt from other general controls such as requirements for post-market reporting.
- *Class II (moderate-risk).* Moderate-risk devices generally require 510(k) review prior to their introduction to the market. Furthermore, a new 510(k) pre-submission is required for Class II devices if a manufacturer/distributor makes changes to the intended use for a device already in commercial distribution and the intended use remains moderate risk, or there is a modification of a device that could affect its safety or effectiveness.
- *Class III (high-risk).* High-risk devices have substantial importance for the prevention of impairment of health, or have a potentially unreasonable risk of illness or injury for which general and special controls are considered inadequate. These types of devices require a premarket approval (PMA) prior to commercial distribution.

A device subject to 510(k) review must be assessed for substantial equivalence to another legally marketed device, called a predicate.[12] This usually includes pre-analytical and analytical data and, in most cases, includes at least data from patient samples with any further necessary clinical studies depending on pre-existing information on the characteristics of the device/analyte combination. In general, it is insufficient for an assay sponsor to demonstrate device performance in its own laboratory because the data to establish the performance of the device should reflect the performance in routine clinical use. Therefore, the FDA will require that certain studies to evaluate assay performance be conducted in representative clinical laboratories that reflect the expected users for the test system.[12]

Depending on the risk, a de novo review process may be pursued when there is no predicate device to which a new device can be found substantially equivalent. This de novo down-classification process is intended to address low or moderate risk devices that have novel intended uses for which there is no legally marketed predicate device. Procedurally, these devices are automatically assigned as class III (PMA), after which the FDA may choose to down-classify them to class I or II, if the risk is sufficiently low based on its intended use. The new device performance is assessed for safety and effectiveness the same way as if it were the subject of a PMA.

As of January 19, 2011, the FDA made public its plans to propose changes to regulatory processes in order to drive innovation and bring important technologies to patients in an expedited fashion. Such changes include the appeal of decisions made by the Center for Devices and Radiological Health, the submission of clinical data, the characterization of "intended use", the identification of safety and effectiveness issues, and the classification system for de novo 510(k) filings (http://www.fda.gov/AboutFDA/CentersOffices/CDRH/CDRHReports/ucm239448.htm). These plans aim to make the 510(k) program a blueprint for smarter medical device oversight. In addition, the FDA intends to put out a number of draft guidances covering the proposed changes in 2011.

MULTIPLEX PROTEIN-BASED IN VITRO DIAGNOSTIC ASSAYS

According to the FDA, an in vitro diagnostic multivariate index assay (IVDMIA) is a device that combines the values of multiple variables (from multiple measurements using multiple tests) using an interpretation function to yield a single, patient-specific result (eg, a classification or score) intended for the diagnosis of diseases, or in the cure, mitigation, treatment, or prevention of diseases, and provides a result whose derivation is nontransparent and cannot be independently derived or verified by the end user (**Fig. 1A**).[13] These types of tests have been developed based on observed correlations between multivariate data and clinical outcome. (**Fig. 2**). The FDA draft guidance on IVDMIA devices is currently available at: http://www.fda.gov/downloads/MedicalDevices/DeviceRegulationandGuidance/GuidanceDocuments/ucm071455.pdf.

As part of the FDA's Critical Path Initiative to medical product development (http://www.fda.gov/ScienceResearch/SpecialTopics/CriticalPathInitiative/default.htm), the MicroArray Quality Control (MAQC) consortium began in February 2005 to address various microarray reliability concerns raised in publications pertaining to reproducibility of gene signatures.[14–16] The first phase of this project (MAQC-I) extensively evaluated the technical performance of microarray platforms in identifying all differentially expressed genes that would potentially serve as biomarkers. This project demonstrated high intraplatform reproducibility across test sites, as well as interplatform concordance of differentially expressed gene lists,[17,18] and also confirmed that microarray technology is able to reliably identify differentially expressed genes between sample classes or populations.[19,20] The second phase project (MAQC-II) addressed the development and evaluation of accurate and reproducible multivariate gene expression-based prediction models, referred to as classifiers and composite scores.[21] For any given microarray dataset, many computational approaches can be followed to develop predictive models and to estimate the future performance of these models. This study found that model performance using different approaches generated models of similar performance. Hence, the conclusions and recommendations from MAQC-II should be useful for regulatory agencies, study committees, and independent investigators that evaluate methods for global gene expression analysis. Additionally, the good modeling practice guidelines established by MAQC-II and the lessons learned from this project provide a solid foundation from which other high-dimensional biological data could be more reliably used for the purpose of predictive and individualized medicine. For proteins, an example of IVDMIA is the recently cleared OVA1 test (Vermillion, Inc), the first blood test based on 5 previously known markers to help a physician evaluate the likelihood that an ovarian mass is malignant or benign prior to a planned surgery (more information available at http://www.accessdata.fda.gov/cdrh_docs/reviews/K081754.pdf).

A multiplex protein assay is defined as a device/test system where one or more protein/peptide targets are simultaneously detected via a common process of sample

A Strategy for Multiplex IVD Development

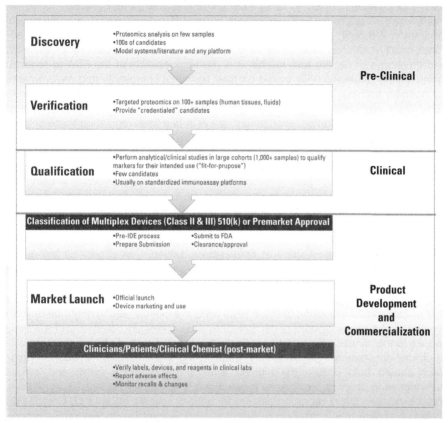

Fig. 2. A strategy for *in vitro* diagnostics (IVD) product development and commercialization of multiplex protein-based assays.

preparation, measurement, and interpretation (**Fig. 1**B).[13] A multiplex assay can either generate a composite "score" using IVDMIA algorithms (behind the dash line) or not (ie, collective assessment of individually measured values from one common test). Examples of FDA-cleared multiplex protein assays include the flow immunoassay for the qualitative detection of IgG antibodies of Epstein-Barr Virus Nuclear Antigen, Viral Capsid Antigen, and Early Antigen diffuse in human serum to aid in the diagnosis of infectious mononucleosis (Bio-Rad Laboratories, Inc, http://www.accessdata.fda. gov/cdrh_docs/reviews/K062211.pdf), and the ENA IgG BeadChip Test System intended for use in testing in human serum the presence of human IgG class antibodies to 6 extractable nuclear antigens (ENAs) (BioArray Solutions, Ltd, http:// www.accessdata.fda.gov/cdrh_docs/reviews/K043067.pdf). Currently, there are no multiplex protein assays cleared by the FDA on a mass spectrometry platform.

FDA's POSITION ON REGULATING MULTIPLEX PROTEIN ASSAYS

Laboratory-developed tests (LDTs or "home-brew tests") were initially either well-characterized, low-risk diagnostics or for rare diseases for which adequate validation would not be feasible and the tests were being used to serve the needs of the local patient population. Currently, LDTs exist for at least 96 protein analytes for which there is no FDA-approved test. Some IVDMIAs are LDTs developed by a single clinical laboratory. Although Clinical Laboratory Improvement Amendments (CLIA) establishes quality-testing standards for all laboratories to ensure the accuracy, reliability, and timeliness of patient test results, it is the responsibility of the FDA to regulate commercially marketed IVDs based on their technical complexity and requirements for operator training. The categories of IVDs include waived tests, tests of moderate complexity, and tests of high complexity. Historically, LDTs have been a subject of FDA enforcement discretion, even though the FDA has been regulating some of them (eg, tests used in blood banking). Additionally, the components of traditional LDTs were regulated individually by FDA as Analyte Specific Reagents (ASRs) or other specific or general reagents, and the tests were (as is still true today) developed and offered in CLIA high-complexity laboratories with extensive experience in using the tests. Currently, diagnostic tests are playing an increasingly vital role in clinical decision-making and disease management, particularly in the context of personalized medicine. As a result, LDTs that have not been properly validated for their intended use may put patients at risk. Therefore, the regulatory agency has indicated an interest in changing its longstanding policy of enforcement discretion as shown by the recent Federal Register notice (http://edocket.access.gpo.gov/2010/2010-14654.htm) and the FDA's public meeting on the oversight of LDTs in July 2010.

In the context of IVDMIA and multiplex protein-based assays, the FDA considers that proper clinical validation of these tests used in a clinical setting might be beyond the scope of a single laboratory because they are intrinsically complex and will likely require further simplification and analytical robustness in order to be used extensively in clinical applications. Consequently, establishing a standardized evaluation paradigm should help ensure the highest level of performance within and across laboratories in order to provide the most accurate results for patients. Furthermore, there is no mandatory reporting system for adverse events (AEs) of LDT results on patient outcome, even though the FDA has encouraged healthcare professionals, consumers, laboratories, manufacturers, and others to voluntarily report AEs and malfunctions associated with LDTs to its MedWatch system (http://www.fda.gov/Safety/MedWatch/default.htm). To provide a clear pathway and encourage increased reporting, a separate product code for these tests is currently being implemented in order to facilitate voluntary reporting of AEs associated with LDTs.

REGULATION OF MULTIPLEX PROTEIN ASSAYS ON MASS SPECTROMETRY PLATFORMS

Mass spectrometers (MS) for clinical use, as described under 21 CFR 862.2860, are considered Class I devices (general controls) and are exempt from the 510(k) process, unless they are used for the diagnosis, monitoring, and screening of neoplastic diseases, cardiovascular diseases, and diabetes (http://www.accessdata.fda.gov/scripts/cdrh/cfdocs/cfcfr/CFRSearch.cfm?fr=862.9). MS under 510(k) exemption are still required to follow GMP guidelines (http://www.fda.gov/MedicalDevices/DeviceRegulationandGuidance). By law, medical device manufacturers must incorporate several elements related to the manufacturing, packing, storage, and installation of devices (to conform to Current Good Manufacturing Practices [cGMPs]) into their quality assurance program in order to meet the requirements of the Quality System Regulation (QSR). Currently, there are 10 MS listed in the 510(k) database under this exempt status.[13] Nevertheless, Class I devices should be registered and listed regardless of whether they require a premarket review. FDA regulations distinguish between these Class I mass spectrometers (cleared between 1977–1999) requiring only general controls, and more complex instrumentation such as those instruments that include complex measurement functions and interpretive software, which may be regulated similarly to the multiplex instrumentation for nucleic acid assays (Class II). Since instrumentation used to run a specific assay takes on the classification of that assay for its intended use, a mass spectrometer submitted as part of a Class III assay would have to be evaluated as a Class III instrument for that intended use.

As previously defined,[13] a multiplex protein-based MRM-MS assay coupled to High Performance Liquid Chromatography (HPLC) simultaneously measures multiple proteolytic peptides via a single HPLC injection. Quantitative MRM-MS measurements on peptides to obtain overall "protein concentrations" in human biospecimens (plasma, serum, proximal fluids, or tissues) are considered multiplex. Since the design of modern LC-MS instruments, including triple quadrupole mass spectrometers used for MRM measurements, can vary between different manufacturers, potentially resulting in nonequivalent performance of assays on different instruments, the submission review would require different specifications for each instrument.[13] In any case, the "pre-Investigational Device Exemption" (pre-IDE) process is recommended as an informal "pre-submission" procedure to engage the FDA in early communication with the sponsor. This may involve sending analytical or clinical protocols to the FDA for review and comment before proceeding with studies, or meeting with the FDA to discuss protocols and/or possible regulatory pathways. This process does not mean that manufacturers are required to subsequently submit an IDE application. In fact, most IVDs are exempt from the medical device IDE regulations.

To begin to assess the feasibility of MRM-MS assays in clinical settings, the NCI-CPTC initiative, through its Clinical Proteomic Technology Assessment for Cancer (CPTAC) network of research centers, conducted an interlaboratory study as a first step to demonstrate analytical reproducibility of this technology across instrument platforms and laboratories. By targeting 11 peptides from 7 proteins using corresponding heavy isotope labeled internal standards as spike-ins at 9 different concentrations in depleted human plasma, the 8 individual labs were able to produce results with CVs ranging between less than 5% to 23%.[22] Even in the most complex scenario, in which multi-step sample preparation was performed at these individual sites to simulate realistic biomarker research, the highest CV was less than equal to 23% using a single transition of MRM for all peptides except for one outlier. These preliminary (preclinical) results show the promise of these kinds of assays in the clinics. Further development in automation to streamline sample preparation, in conjunction with software development, should eventually reduce labor-intensive workflow, analytical variability between different instrument platforms, and the need for a high level of expertise currently required to perform this kind of assays. Once analytical performance can be demonstrated as equivalent on different MS platforms, separate regulatory evaluations of the same analyte(s) on each instrument may become simplified or unnecessary.[13]

To satisfy FDA requirements, performance data using a specific MS platform in combination with any other instrumentation as a part of the test system (eg, pre-analytical, HPLC) to perform a specific assay should be provided as a part of the initial submission for an assay. If performance is deemed adequate, this generally leads to an approval of an assay only on that specific instrument used in evaluating performance, as the assay performance on other platforms would be unknown. This can be followed by subsequent submissions addressing any modifications to the assay for the same intended use (see above; eg, an addition of another instrument platform). This would make that instrument manufactured under a quality system available for clinical laboratory testing. Any changes to this cleared platform, including all necessary components and accessories (eg, HPLC columns, traps, tubings, electrospray source, software), should be re-evaluated by the assay manufacturer, and may or may not need subsequent regulatory submissions to address the safety and effectiveness of the changed platform. Additional analytes or kits can subsequently be cleared for use on the same platform without additional platform-specific information, other than assay-specific components of the instrumentation.

ANALYTICAL AND CLINICAL VALIDATION CONSIDERATIONS

A premarket submission document should include: a device's intended use/indication for use, a description of the device covering both the instrument and reagents, and analytical and clinical performance studies evaluating performance of the device for its intended use. Device clearance or approval rests on the ability of the sponsor to provide analytical and clinical data that demonstrate adequate performance of the device for its claimed intended use. The extent of the required data will largely depend on whether the markers measured have been previously established as clinically useful for the intended use of the assay. In case of novel markers, simple analytical detection or quantification of an analyte is inadequate. Significance of the measurement of novel markers for clinical management of the patient must also be demonstrated, either through clinical data or, in some cases, through sufficient credible published information supporting clinical use.

In the analytical performance section of a premarket submission, the performance of the device should be clearly described in terms of precision, accuracy, and performance around the cut-off point, along with other performance measurements such as specificity, sensitivity, linearity, limit of detection, and limit of quantitation, as applicable. A detailed description of appropriate internal and external controls and calibrators used in the assay should also be included for evaluation. It is important for the sponsor to demonstrate that all protein analytes in a panel must meet the FDA's performance criteria rather than extrapolating the performance of one analyte to the others. Furthermore, a multiplex MRM-MS protein assay should carefully address cross-reactivity (if any) or interference of analytes within and outside the panel.

With regard to internal and external controls required for the submission, the quantitation of proteins in plasma/serum, based on multiplex MRM-MS measurements of peptides, should consider the following: controls for assessing the efficiency and variability of proteolysis in different samples, controls for gauging analytical recovery of proteins/peptides and/or post-translational modifications (PTMs) during sample preparation that involves an immunoaffinity enrichment step.[23] In any case, the reproducibility of the "overall measured protein concentration" from target peptides derived from the same protein would need to be demonstrated as consistent among replicates. If one of the peptide measurements is consistently an outlier, sponsors need to understand why this occurs and what effect it would have on the assay, and if the outlier and its effect would be recognized by the assay QC system as an outlier which could signal the presence of PTMs, interferences from other proteins, single nucleotide polymorphisms (SNPs), and so forth.

An additional key consideration for IVDMIAs or multiplex protein assays is the use of interpretive software to reach a patient-specific result (risk score or classifier) to aid in clinical decision-making.[24,25] The FDA generally requires that software algorithms included in the assay for data and results interpretation be pre-specified before analyzing study data. Alteration of the algorithm to better fit the data after the study is performed is generally unacceptable.

CLINICAL CHEMISTS

The Clinical Laboratory Improvement Amendments require clinical labs to perform analytical validation for all tests prior to implementation. For FDA-cleared tests, this process involves post-market verification of the manufacturers' claims for the performance specifications, which helps to ensure that a test, when used in a clinical laboratory by testing personnel for its patient population, is performing as the manufacturer intended. This post-market verification process (not "verification" stage in the NCI-CPTC biomarker pipeline) is performed as QC procedures to evaluate whether a previously cleared test (eg, instrument, controls, reagents) complies with regulations, specifications, or conditions. For non-FDA-cleared tests, labs must establish the criteria for each of the performance specifications as a part of their analytical validation procedures. Although CLIA regulations do not specifically address clinical qualification, clinical labs are held responsible for all aspects of their operation, including test selection. Clinical chemists routinely validate analytical performance characteristics such as imprecision, bias, and the limits of quantification necessary to detect analytes in human samples. Since the methods of proteomics are presently more complex and less defined than traditional clinical chemistry methods, clinical chemists have an opportunity to play a role in educating basic researchers on the importance of these performance characteristics and the proper way to assess these characteristics. For instance, in verifying (QC) an IVDMIA and/or a multiplex protein assay using a "score" or "classifier", a clinical lab should design an approach which may involve running an adequate number of positive and negative patients to assess the performance of the "score" in their diagnosis in comparison to their medical charts,and the final clinical diagnosis to ensure the accuracy of the software algorithm.

CLINICAL LABORATORY STANDARDS INSTITUTE

The Clinical Laboratory Standards Institute (CLSI) is highly recognized and respected by organizations of clinical professionals. The CLSI provides useful information to sponsors and the FDA in the process of preparing and reviewing 510(k) and PMA submissions since CLSI documents go through rounds of rigorous review prior to publication (http://www.clsi.org). These documents are developed and approved by consensus of stakeholders in particular areas and go through public comment phase. The FDA can either fully or partially recognize CLSI documents as standards, and compliance with the recommendations of CLSI documents may be accepted as evidence of fulfillment of certain FDA analytical requirements. In the event that the FDA does not fully recognize a CLSI document, it can still be referenced in the submission. One such useful document, EP-17A, Vol. 24, No. 34, Protocols for Determination of Limits of Detection and Limits of Quantitation; Approved Guideline, is commonly referenced for quantitative assays and can be very informative for proteomic researchers developing multiplex MRM-MS assays. While there is currently no CLSI guidance documents for multiplex protein assays, general information could be drawn from the nucleic acid-based multiplex world (eg, MM-17A;Verification and Validation of Multiplex Nucleic Acid Assays; Approved Guideline).

REFERENCES

1. Polanski M, Anderson L. A list of candidate cancer biomarkers for targeted proteomics. Biomark Insights 2007;1:1–48.
2. Anderson NL. The clinical plasma proteome: a survey of clinical assays for proteins in plasma and serum. Clin Chem 2010;56:177–85.

3. Hortin GL, Carr SA, Anderson NL. Introduction: advances in protein analysis for the clinical laboratory. Clin Chem 2009;56:149–51.

4. García-Foncillas J, Bandrés E, Zárate R, et al. Proteomic analysis in cancer research: potential application in clinical use. Clin Transl Oncol 2006;8:250–61.

5. Cairns DA. Statistical issues in quality control of proteomic analyses: Good experimental design and planning. Proteomics 2011. [Epub ahead of print].

6. Paulovich AG, Whiteaker JR, Hoofnagle AN, et al. The interface between biomarker discovery and clinical validation: The tar pit of the protein biomarker pipeline. Proteomics Clin Appl 2008;2:1386–402.

7. Regnier FE, Skates SJ, Mesri M, et al. Protein-based multiplex assays: mock presubmissions to the US Food and Drug Administration. Clin Chem 2010;56: 165–71.

8. Whiteaker JR, Zhao L, Anderson L, et al. An automated and multiplexed method for high throughput peptide immunoaffinity enrichment and multiple reaction monitoring mass spectrometry-based quantification of protein biomarkers. Mol Cell Proteomics 2010;9:184–96.

9. Keshishian H, Addona T, Burgess M, et al. Quantification of cardiovascular biomarkers in patient plasma by targeted mass spectrometry and stable isotope dilution. Mol Cell Proteomics 2009;8:2339–49.

10. Anderson NL, Jackson A, Smith D, et al. SISCAPA peptide enrichment on magnetic beads using an in-line bead trap device. Mol Cell Proteomics 2009;8: 995–1005.

11. Zhao M, Nolte D, Cho W, et al. High-speed interferometric detection of label-free immunoassays on the biological compact disc. Clin Chem 2006;52:2135–40.

12. Mansfield E, O'Leary TJ, Gutman SI. Food and Drug Administration regulation of in vitro diagnostic devices. J Mol Diagn 2005;7:2–7.

13. Boja ES, Jortani SA, Ritchie J, et al. The Journal to Regulation of Protein-Based Multiplex Quantitative Assays. Clin Chem 2011;57:560–7. [Epub ahead of print].

14. Frantz S. An array of problems. Nat Rev Drug Discov 2005;4:362–3.

15. Shi L, Tong W, Goodsaid F, et al. QA/QC: challenges and pitfalls facing the microarray community and regulatory agencies. Expert Rev Mol Diagn. 2004;4:761–77.

16. Michiels S, Koscielny S, Hill C. Prediction of cancer outcome with microarrays: a multiple random validation strategy. Lancet 2005;365:488–92.

17. Shi L, Reid LH, Jones WD, et al. The MicroArray Quality Control (MAQC) project shows inter- and intraplatform reproducibility of gene expression measurements. Nat Biotechnol 2006;24:1151–61.

18. Patterson TA, Lobenhofer EK, Fulmer-Smentek SB, et al. Performance comparison of one-color and two-color platforms within the MicroArray Quality Control (MAQC) project. Nat Biotechnol 2006;24:1140–50.

19. Irizarry RA, Warren D, Spencer F, et al. Multiple-laboratory comparison of microarray platforms. Nat Methods 2005;2:345–50.

20. Strauss E. Arrays of hope. Cell 2006;127:657–9.

21. Shi L, Reid LH, Jones WD, et al. The MicroArray Quality Control (MAQC)-II study of common practices for the development and validation of microarray-based predictive models. Nat Biotechnol 2010;28:827–38.

22. Addona TA, Abbatiello SE, Schilling B, et al. A multi-site assessment of precision and reproducibility of multiple reaction monitoring-based measurements by the NCI-CPTAC network: toward quantitative protein biomarker verification in human plasma. Nat Biotechnol 2009;27:633–41.

23. Ciccimaro E, Hanks SK, Yu, KH, et al. Absolute quantification of phosphorylation on the kinase activation loop of cellular focal adhesion kinase by stable isotope dilution liquid chromatography/mass spectrometry. Anal Chem 2009;81:3304–13.
24. Russek-Cohen E, Tezak Z, Philip R, et al. Multivariate Assays Reporting a Composite Score. JSM Proceedings CD-Rom. 2007.
25. Rodriguez H, Tezak Z, Mesri M, et al. Analytical validation of protein-based multiplex assays: a workshop report by the NCI-FDA Interagency Oncology Task Force on Molecular Diagnostics. Clin Chem 2010;56:237–43.

Erratum

Clin Lab Med 2011; 31; 359

A Retraction printed in the June 2011 issue of Clinics in Laboratory Medicine should have referred to the following article: From Stem Cell to Red Blood Cells In Vitro: The 12 Labors of Hercules Clin Lab Med 2010; 30:391-403.

Clin Lab Med 31 (2011) 455
doi:10.1016/j.cll.2011.06.001
0272-2712/11/$ – see front matter

Index

Printed and bound by CPI Group (UK) Ltd, Croydon, CR0 4YY

03/10/2024

01040455-0012